CONTESTED ENVIRONMENTALISMS

Contested Environmentalisms

Trees and the Making of Modern China

CHENG LI

STANFORD UNIVERSITY PRESS
Stanford, California

Stanford University Press
Stanford, California

Printed and bound by CPI Group (UK) Ltd, Croydon, CR0 4YY

Library of Congress Cataloging-in-Publication Data
Names: Li, Cheng (Professor of Chinese Studies), author.
Title: Contested environmentalisms : trees and the making of modern China / Cheng Li.
Description: Stanford, California : Stanford University Press, 2025. | Includes bibliographical references and index.
Identifiers: LCCN 2024013704 (print) | LCCN 2024013705 (ebook) | ISBN 9781503640306 (cloth) | ISBN 9781503641334 (ebook)
Subjects: LCSH: Forest policy—China—History. | Tree planting—China—History. | Forest conservation—China—History. | Forests and forestry—China—History. | Environmentalism—China—History.
Classification: LCC SD643 .L473 2025 (print) | LCC SD643 (ebook) | DDC 634.9/280951—dc23/eng/20240617
LC record available at https://lccn.loc.gov/2024013704
LC ebook record available at https://lccn.loc.gov/2024013705

Cover design: Lindy Kasler
Cover art: Pang Jun, "Youngsters! Use both your own hands to make our great motherland greener," April 1956, 53.5 x 69 cm. BG E12/754, chineseposters.net, IISH collection.

For trees that nurture all of us and our planet, Earth

CONTENTS

LIST OF ILLUSTRATIONS

ACKNOWLEDGMENTS

This book is dedicated to those who believe in my research and in me. I am deeply indebted to a vast number of gifted and supportive colleagues and friends. Beyond doubt, this book would not have been possible without the generous support I received from Jing Tsu at Yale University. Her insight into my project, flexibility regarding my research methods, and warmest encouragement inspired me to enthusiastically explore Chinese environmental ideas. At many critical junctures, her guidance was invaluable and decisive. Paul Sabin's writing seminar on environmental history not only pushed me to consider Chinese environmental issues from a historical perspective but also made me more confident that I could complete this project. His suggestions were always timely and productive. Michael Warner's seminar on American environmental literature led me to enjoy the magic and magnificence of close reading. His insight into literary studies was always eye opening. My discussions with Peter Perdue on Chinese environmental history were enormously encouraging and inspirational. Kang-I Sun Chang trained me to be more proficient in reading classical Chinese literature and provided unyielding spiritual support at several pivotal moments. Cajetan Iheka generously offered constructive and insightful guidance on revisions.

At Yale, I also had the precious opportunity to work with many other dedicated scholars, including Michelle Bell, Luke Bender, Hsiu-hsien Chan, Min Chen, Joe Cleary, Deborah Coen, Aaron Gerow, John Grim, Robert Harms, Denise Ho, Mick Hunter, Edward Kamens, Ninghui Liang,

Fan Liu, Eric Rutkow, James Scott, Jianhua Shen, Kalyanakrishnan Sivaramakrishnan, Mary Tucker, and Yongtao Zhang. I thank Bo An, Yun Bai, Baiqian Bian, Cheow Thia Chan, Po-hsi Chen, Lynna Dhanani, Abigail Fields, Yufang Gao, Emily Jin, Lav Kanoi, Caitlin Kossmann, Yang Hua, John Lee, Carlos Nugent, Rundong Ning, Taylor Rose, Loren Waller, Mengxiao Wang, and Zekun Zhang for their kind support. Dihao Zhou is the most ideal roommate and cook, an exceptional scholar, and the perfect soulmate for my journey at Yale. Sophia Richardson in the English Department provided constructive suggestions on every chapter of my work, for which I am deeply appreciative. Shu-han Luo patiently and wisely guided me in becoming a more compelling author.

My academic journey in the United States started at the University of Wisconsin–Madison, where I met a number of generous and kind professors. Nicole Huang warmly encouraged me in the tough transition from Chinese to American academia during my first two years in the United States. Without her unyielding support and insightful guidance, I would not have survived at UW–Madison. I took my first course on environmental humanities with Rob Nixon, who was always generous, engaging, and encouraging, especially when I was struggling to start my academic journey as a first-year, international student. The discussions in Judd Kinzley's seminar on East Asian environmental history provided me with the opportunity to reevaluate literature from a historical viewpoint. This experience was instrumental in establishing the interdisciplinary approach taken in this book. Steve Ridgely inspired me to view China through a broader trans-Asian perspective, and his insights have continued to motivate me since I became a professor myself. Rania Huntington's enthusiastic support offered me the chance to explore modern Chinese literature from the perspective of classical Chinese texts, and she graciously extended help during a pivotal moment in my graduate studies. Anne McClintock warmly encouraged me during one of the most challenging moments in my life, when I almost abandoned my academic journey. I am also deeply grateful to Gregg Mitman, Viren Murthy, William Cronon, Preeti Chopra, and Susan Friedman. During my brief period of study at the University of Chicago, I had the opportunity to learn from Jacob Eyferth, Fredrik Jonsson, and Kenneth Pomeranz, all of whom enhanced my training in historical methodology. I am grateful for their guidance. Since I was studying in Beijing, Scott Slovic encouraged me to explore ecocriticism and offered timely and constructive suggestions on environmental humanities, for which I extend my most sincere gratitude.

For providing appropriate resources and an encouraging working environment, I must thank Stephen Brockmann, at Carnegie Mellon University, who has read every chapter of the manuscript and offered insightful comments. He is always very patient and knowledgeable in providing constructive answers to my questions. My department head, Anne Lambright, motivated me to start the book revision and submission process. I am grateful to my other colleagues at Carnegie Mellon—Uju Anya, Katharine Burns, Stephan Caspar, Neil Donahue, Sébastien Dubreil, Kenya Méndez, Tatyana Gershkovich, Gang Liu, Christopher Lowy, Abigail Owen, Edmund Russell, John Soluri, Therese Tardio, Zhongxin Sun, Joel Tarr, Noah Theriault, Benno Weiner, Thomas Werner, Seth Wiener, Suemei Wu, Tianxue Yao, Yoshihiro Yasuhara, and Bonnie Youngs—for their dedication to creating a nurturing academic environment. I am also immensely grateful to Kun Qian at the University of Pittsburgh, who provided informative and constructive suggestions on several pivotal chapters.

I would like to express my thanks for the generous financial support from Mellon Area and International Studies Fellowship from the University of Wisconsin–Madison; the Council on East Asian Studies from Yale, including an East Asian Studies Prize Fellowship and two summer research grants; Yale's Beinecke Rare Book and Manuscript Library summer research grant; and the Falk Fund for Research in the Humanities from Carnegie Mellon University. They offered me much-needed free space to pursue my passions. Jie Li encouraged me to expand chapter 2 during a Harvard-Yale graduate student conference in 2019. During the Yale East Asian Studies Prize Fellowship exam in 2020, Karen Thornber and David Der-wei Wang offered valuable guidance and constructive feedback that greatly contributed to the early stages of the book's development. Ban Wang, Sheldon Lu, Micah Muscolino, Robin Visser, Shellen Xiao Wu, and Ling Zhang provided warm encouragement on this project during discussions at the Association of Asian Studies Conference in 2023. During my field research in 2018 and 2019, Fei Li generously shared his informative insights into this project and confirmed many of my early hypotheses. I also want to extend my thanks to other colleagues, teachers, and friends, including David Bello, Yin Cai, Fanghao Chen, Iza Ding, Haomin Gong, Peng Hai, Dingru Huang, Fangyuan Huang, Kun Huang, Sujie Li, Yanjun Liu, Kyhl Lyndgaard, Nan Ma, Bozhou Men, Benjamin Morgan, Maria Romanova, Guldana Salimjan, Mingwei Song, Ying Jia Tan, Ying Tao, Xiqun Wang, Guanda Wu, Lidong Xiang, Lixin Xu, Panpan Yang, Haihui Zhang, Yunqiu Zheng, Chenshu Zhou, Yunjun Zhou, and Runxiao Zhu. They provided useful and professional suggestions and spiritual support, for which I am grateful.

I would also like to extend my gratitude to the East Asian Library at Yale University. Michael Meng's prompt and enthusiastic help with Chinese sources significantly bolstered my confidence in this project; at times when I doubted its feasibility, his insightful guidance convinced me of its practicality. I cannot thank him enough. James Tierney patiently guided me through the pitfalls of English writing. Philip Dematteis provided professional, constructive, and timely copyediting assistance during the pivotal moments of the book's development. His invaluable support was crucial in helping me to complete the book on schedule. ChatGPT 4.0 was consulted to polish the grammar of a very small portion of the book in the final stage of the manuscript. For library support of various types of materials and inquires, I thank Beijing Forestry University, the Chinese Academy of Forestry, Cornell University, the Library of Congress, Peking University, the National Library of China in Beijing, Northeast Forestry University (Harbin, China), the University of California, Berkeley, the University of the Pacific, the University of Michigan, and Yale University Library. At Stanford University Press, I am extremely thankful to Dylan Kyung-lim White, who gave me steadfast support from the book proposal to the review stage, and my anonymous readers for their valuable feedback that improved this book.

A version of chapter 2 appeared in the journal *Environmental History*, published by the University of Chicago Press. An early version of chapter 3 appeared in the *Journal of Chinese Cinemas*. I am grateful to the journals' editors, including Stephen Brain, Mark Hersey, Sheldon Lu, and Song Hwee Lim, for their patient and professional assistance. My thanks also go to the publishers of those journals for allowing me to use those articles in my book. I thank my hardworking parents, Li Zhengfu and Qu Shiqin, for their unwavering support and sacrifices that have paved the way for my achievements. My brother, Li Lun, for his guidance since I was a child, always being a source of wisdom and strength. I am grateful to my beloved spouse, Bill Godfrey, for all the joy we share and the encouragement through all the challenging years. From my academic journey in Siping, Beijing, Madison, Chicago, and New Haven to landing a job in Pittsburgh, I hope my path as a first-generation college student, an immigrant, and the child of oilfield workers has made all of you proud.

I must, finally, thank the trees in the forests of the Catskills in rural upstate New York. During the COVID-19 pandemic, they provided the greatest relief both during and after my intensive writing schedule, offering spiritual support in that unprecedented era.

This book is about trees but not merely about trees. By using trees as an entry point, it explores how the notions of trees participate in nation building and subject formation. Analyzing this interactive process demonstrates the conflicting nature of modern Chinese conservationist ideas. This book is about the environmental notions of China, but it goes beyond China. Unpacking the nuances of Chinese environmental ideas, this book makes a case for the broader implications of global authoritarian environmentalism.

The book begins with the pioneering modern Chinese forester, Ling Daoyang, who came to Yale to study in the 1910s. When I touched his signature and writings at the Yale Manuscript Library in 2018, I leaped out of my chair, deeply emotional and thrilled. It is these pioneering foresters who compelled me to trace the development of modern Chinese environmental ideas over the course of a century. Let us fasten our seat belts and embark on an odyssey through the lush landscape of Chinese environmental ideas.

Once again, I thank everyone who trusted me and gave me hope during the darkest moments of my academic career. All mistakes are mine, but I owe all successes to you.

Introduction

FOR DECADES, TREE PLANTING and forest management have been at the heart of Chinese environmental endeavors.[1] These efforts currently serve as a major part of China's ambition to achieve carbon neutrality by 2060. By employing trees to combat climate change, China—the largest emitter of greenhouse gases in the world—is playing the environmental card to improve its relationships with other global powers. The Chinese approach to trees and forestry is pivotal to its environmentalism and its green image more generally.

The implications of tree planting extend far beyond the physical action itself. It is impossible to understand modern Chinese environmentalism without knowing how trees have shaped Chinese society. "Planting trees is planting happiness," reads a faded 2019 billboard near the subway entrance of the National Library of China in Beijing (Figure 0.1). Listed alongside the "socialist core values" of prosperity, democracy, civilization, harmony, and equality, tree planting is depicted as an essential step toward realizing these values. Associating this host of virtues with flourishing trees, the government strives to weave tree planting and its associated cultural values into the Chinese mind.

Contemporary Chinese tree-planting efforts are the culmination of a century-long saga of incorporating forestry into Chinese nation building and cultivating active citizenship. A special issue of the prominent Shanghai newspaper *News* (*Xinwen bao*), published during the city's celebration

FIGURE 0.1 **"Planting trees is planting happiness," reads a billboard in the subway entrance of the National Library of China in Beijing in August 2019. It is listed alongside the "socialist core values" of prosperity, democracy, civilization, harmony, and equality. Riding a bike, like the one parked near the billboard, also contributes to tree planting thanks to the *Ant Forest* project by China's e-commerce giant Alibaba. Photo by author.**

of Arbor Day on March 12, 1928, uses a wide-ranging collection of slogans to elaborate on how trees could affect almost every aspect of citizens' lives:

Trees are the essays written by land and the art of nature.

Trees are citizens' companions and the soul of the city.

Trees turn smog-filled Shanghai into a tree-lined city.

Planting trees can improve the beauty of the streets.

Planting trees is the solution to improving citizens' boring lives.

Planting trees is a first step toward the ruralization of a city.

Planting trees will provide a foundation for the Sun Yat-sen Forest.

Planting trees is the best way to commemorate Sun Yat-sen.

Emphasizing tree planting began in the era of political tutelage.

Planting trees can mitigate flood and drought.

Planting trees can help produce the raw materials for industry.

Planting trees is the most solid method of depositing.

Trees can temper the climate and improve hygiene.

All citizens of a civilized country protect public trees.

Forests are a natural school for children.

Trees are a living textbook for the study of nature.

Protecting trees is the responsibility of citizens.

Tree planting becomes truly valuable only when it becomes a public movement.[2]

The slogans, together with other articles in the issue, called on Shanghai residents to participate in the tree-planting movement at a time when the Chinese economy was collapsing, military and political conflicts were escalating, and environmental disasters abounded. The slogans promoted tree planting as a solution—even a panacea—for various urgent social problems: trees can yield economic profit; prevent ecological disasters; improve the aesthetics of the city; memorialize Sun Yat-sen, the founding father of Republican China; enhance citizens' mental health; and educate youth.

Read closely, however, the slogans reveal an inherent tension in promoting *afforestation* (establishing forests on previously unforested land). On the one hand, they advocate planting trees for their ecological, psychological, and aesthetic benefits. On the other hand, they suggest that trees should be harvested, thereby providing timber for industrial development to boost the ailing economy and improve the welfare of impoverished citizens.[3] From the founding of Arbor Day in 1915 and the inauguration of the tree-planting movement in the 1920s through the Greening the Motherland Campaign in the 1950s to the building of the Three-North Shelter Forest (Sanbei fanghulin) since the 1980s, China has carried on a century-long afforestation initiative to popularize environmental notions and solve ecological problems. Nevertheless, there has also been "three great cuttings"—during the Great Leap Forward (1958–60), the Cultural Revolution (1966–76), and the mid-1980s—that devastated China's forests.

The insertion of tree-cutting rhetoric into a conservationist agenda embodies one of the striking tensions in modern Chinese environmental ideas. Investigating these tensions is the central focus of *Contested Environmentalisms*. Through an examination of the conceptual and cultural dynamics of trees and forests, this book explores how conservationist ideas in modern China are inherently in conflict, providing both promises and pitfalls. In twentieth-century China, modern conceptions regarding the benefits of trees and conservationist ideas associated with these conceptions began to circulate among the masses and played a significant role in shaping Chinese identity, structuring daily activities, and promoting active citizenship. This book traces the captivating and convoluted journey of Chinese conservationist attitudes.

Contested Environmentalisms

The central claim of this book is that modern Chinese ideas about trees and conservation are highly contested, revealing the multivalence and malleability of Chinese environmentalism in appealing to the Chinese people. The various tensions—whether to plant trees as a real practice or a symbolic gesture (chapter 1), to interpret nature as static or dynamic (chapter 2), to envision knowledge of nature as universal or as a localized ethnic experience (chapter 3), to understand nature in romantic or militaristic terms (chapter 4), to attempt to resolve China's environmental disasters using native or foreign traditions (chapter 5), and whether to take a socialist or capitalist approach in grappling with these disasters in postsocialist China (coda)—constitute the engine of this book.

Contested Environmentalisms contributes to both modern Chinese studies and environmental humanities in three respects: (1) it shows how trees and afforestation initiatives have shaped nationhood and citizenship through the integration of conservationist discourse into nation building in China; (2) by using trees and their surrounding discourse as a lens to examine modern China, it shows the plurality and, more importantly, the contestation of environmentalisms in China through the circulation of conservationist rhetoric; and (3) it historicizes Chinese authoritarian environmentalism and provides a comparative perspective on global environmentalism.

This book demonstrates how trees and forests shape almost all aspects of life in China, deeply affecting political psychology and legitimacy (chapter 1), scientific reasoning (chapters 2 and 5), ethnic identity (chapter 3), youth development and military mobilization (chapter 4), intellectual movements (chapter 5), and the technological enterprise (coda). The implications of tree planting far exceed the actions themselves. In the eyes of politicians, planting a tree compels a person to cultivate a scientific worldview, develop a forward-looking and optimistic vision, and express loyalty to the nation-state. For businessmen, tree planting integrates their pursuit of profit with their ethical responsibilities to disadvantaged regions. For intellectuals, tree planting drives home the message of embracing Western science and its associated political ideology based on democratic participation. Tree planting's apparent conservationist intentions usually camouflage its opaque but more enduring cultural and political impacts.

I claim that contestation is a defining property of modern Chinese environmentalism. While contestation may seem to be an inherently negative characteristic, it is actually a strength. Contestation endows Chinese environmentalism with its vitality and tenacity. Shuttling between two competing or even contradictory ends, Chinese conservationist ideas make themselves available to an ever-broader audience. Because of this flexibility, contested Chinese conservationist ideas have even contributed to the resilience of Chinese governance insofar as officials have been able to adapt their strategies to support shifting social and political agendas. The coexistence and clash of multiple environmental notions contribute to a rich cultural repertoire that can be evoked and reinvented for future generations. These conflicting varieties of modern Chinese environmentalism function as an engine, thus propelling Chinese environmentalism in an ever-evolving project.

How does the contestation of environmentalism take place in modern China? I argue that the tension between considering tree-planting ideas as universal truths, on the one hand, and as a fluid and constantly adaptable process, on the other hand, gives rise to the contestations of environmental discourse. Modern Chinese environmentalism is thus neither simply a destructive mindset nor a wholehearted embrace of afforestation and conservation. When disparate groups deploy divergent environmental ideas, these ideas respond, thrive, and persist as they accommodate changing social situations.[4] Though their ideas are contested and even contradictory, each group may actually represent part of the truth. However universal they claim their environmental ideas to be, the ideas are true *only* in a certain time and space.[5] Without a specific context, the explanatory power of these putatively universal ideas collapses.

In practice, each party reframes its narrative to achieve its own preconceived goals in the name of the universal. The actual applications manifest themselves not only in a contentious and coercive manner, as in the Chinese borderlands, but also peacefully, among the environmental scientists themselves. In the borderlands, the message of the presumed benefits of trees is pressed upon farmers and ethnic minorities but encounters resistance as it conflicts with their own experiences and frameworks for dealing with the natural world; some inhabitants, for example, consider trees invasive species. Imbued with religious and ethnic connotations, local understandings of desert and afforestation stand in sharp contrast to the universal-conservation rhetoric endorsed by environmental scientists (see chapter 3). On the other hand, the scientists themselves have failed to reach a consensus on the optimal means of conserving the natural world, and they contradict one another on significant issues (as discussed in chapter 2). These conflicts result not only from individuals' economic, social, and political agendas but also from their personal experiences and professional training. Recognizing these heterogeneous origins and disparate political, economic, and cultural dynamics should contribute to an ethics of mutual respect, humility, and tolerance.

The widespread appeal of trees and the natural world functions as a rhetorical bank that can be drawn upon to meet various cultural and political needs: trees operate as an enduring open-ended discourse that can be transformed and channeled in different directions. While conservation discourse appears to be a prevailing and universal concept, its consumption may encounter local variations, hesitations, and struggles.[6] In 1980s China, conservation rhetoric led to profound change when it began to threaten the legitimacy of China's political structure. Published just before the Tiananmen incident in 1989, Dai Qing's edited volume *Yangtze, Yangtze* (*Changjiang, Changjiang*) opposes the construction of the Three Gorges Dam and criticizes the government's lack of openness to environmental management and its suppression of dissident opinions.[7] Environmental discourse, when invoked, thus can transcend the environmental dimension and unleash a transformative and even revolutionary power.

The circulation of conservationist ideas in China also has a transnational dimension.[8] In the 1910s and 1920s, Chinese foresters educated in the United States and Japan promoted forestry as a serious academic discipline and integrated forestry into China's national economic and social fabric;[9] tree planting was transformed into a symbol of political legitimacy, peace, and confidence. During the Mao era, Chinese foresters and environmental scientists were indebted to those in the United States and Germany; they extolled the benefits of environmental management for the public welfare and integrated these benefits with revolutionary rhetoric. The transnational vision provides a universal scientific and technological currency that allows politicians, businesspeople, and scientists to camouflage their political goals of social engineering beneath a veneer of environmental conservation.[10] Planting trees not only plants nature (seemingly universal) but also implants values (locally defined).

From a temporal perspective, the generation of these cultural values is a dynamic, robust, and adaptive process, resulting in the coexistence of heterogeneous values and the lack of a single dominant environmentalism in modern China. Raymond Williams identifies a significant feature of this process of change: "There is indeed regular confusion between the locally residual (as a form of resistance to incorporation) and the generally emergent."[11] Older environmental ideas do not vanish; new ones emerge and may contradict the earlier ones. Because of this complex and dynamic process, Williams argues, "no mode of production and therefore no dominant social order and therefore no dominant culture ever in reality includes or exhausts all human practice, human energy, and human intention."[12] The following chapters scrutinize these contested forms of modern Chinese conservationist ideas—real or symbolic, interpreting nature as static or dynamic, envisioning it on both micro- and macroscales, understanding nature as romantic or militarized, employing native or foreign traditions to address ecological threats, and evoking either a capitalist profit-seeking or socialist grassroots approach to grappling with natural disasters.

Contested Environmentalisms also suggests the limits of advancing conservationist attitudes while they are politicized. To acquire more purchase, environmental ideas have had to gradually adapt to new political and social contexts while finding a way to avoid being marginalized during the breathtaking changes that have been occurring in modern China. Because of their ability to accommodate themselves to changing and even chaotic sociopolitical conditions, modern Chinese environmental ideas have survived and thrived. But while these conflicting notions offer diverse platforms for environmental ideas to be disseminated across various groups, the original ideas that promoted environmental conservation are being compromised. This process reveals a paradox: the more conservationist ideas in China merge with mainstream political agendas, the greater the risk they face of being marginalized.

Uncovering Chinese Conservationist Ideas: An Interdisciplinary Dialogue

A corollary argument of this book is that it is the foresters and environmental scientists of the early and mid-twentieth century, rather than the canonical modern Chinese literary writers, who strove to disseminate conservationist ideas to the general public and have contributed the most to the rise of conservationist attitudes in Chinese literature and culture. Without the writings of foresters and environmental scientists—including scientific works, newspapers, and political propaganda—environmental conservation would not have taken root among the masses. These writings form the core of modern Chinese conservationist representations, which led to the emergence of more explicit conservationist notions in Chinese literature and culture around the 1980s.

This book's investigation of modern Chinese conservationist ideas proposes a paradigm shift of focus within the field of Chinese environmental humanities from environmental actions to environmental ideas, from literary representations to underlying environmental attitudes: it seeks to investigate the environmental attitudes that propel both literary writing and historical actions. Literary scholars such as Karen Thornber, Ban Wang, Robin Visser, and Corey Byrnes explore literary representations of the environment in modern Chinese literature by looking at, respectively, the ambiguity of environmental ideas and actions, the interaction between labor and ecology, the dynamics of ethnicity and nature in the borderland, and the interplay between the imaginary and the material.[13] *Contested Environmentalisms* builds on Thornber's pioneering *Ecoambiguity: Environmental Crises and East Asian Literatures,* which uses literary texts to reexamine the conventional romantic interpretations of modern Chinese literature and culture. Drawing on more diverse forms of media and focusing on critical historical moments in modern China, the present book takes the ambiguity and dynamism of modern China as its premise and seeks to account for various forms of environmental attitudes held by politicians, scientists, ethnic groups, literary authors, filmmakers, and even cellphone-app developers across a century. Whereas Thornber prioritizes the commonality of ecological ambiguity in Chinese literature and culture across East Asian literatures, *Contested Environmentalisms* highlights the distinctive modes of contestation of environmental attitudes in different decades and various forms of media in a Chinese society undergoing a process of drastic change. *Contested Environmentalisms* extends the latest research into the role of the environment in shaping national identity, incorporating perspectives from both Marxist and non-Han ethnic viewpoints as presented by B. Wang and Visser.[14] Exploring literary writings on the Three Gorges Dam, Byrnes suggests that poetry and other works of literary imagination have become active agents in propelling material changes on the Yangtze River.[15] I would suggest that we reverse this perspective by asking how natural agents, particularly trees—emblematic of conservation—contribute to literary representations. By probing the varied rhetorics used to understand trees, this book explores the historical dynamics of the generation, evolution, and even devolution of environmental ideas that have given rise to those representations, thereby illuminating the historicity of environmental attitudes.[16]

Going beyond the level of literary representations, this book focuses on how conservationist attitudes and writings shape attitudes among the general public. While quite a number of these works usually reflect either an individualist-oriented, implicit, or metaphorical sense of environmental protection, this book focuses on a broader range of writings and other media that convey modern Chinese conservationist ideas, especially those circulating throughout modern China and affecting the masses. These writings and media explicitly address imminent ecological degradation to a broader spectrum of readers and audiences. Literary works of Chinese conservation are the crystallization of lively minds and varied historical agents across various genres over the course of a century.

Contested Environmentalisms also aspires to initiate a conversation with the field of environmental history, which has a tradition of presenting a narrative of ecological decline.[17] Recent scholarship on Chinese environmental history has aimed to revise this narrative, invoking Chinese proconservation policies and attitudes.[18] This line of scholarship offers a more thorough and nuanced picture of Chinese environmental changes.

Instead of either regarding Chinese environmental history simply as evidence of a destructive mindset or, on the contrary, painting a rosy picture of Chinese afforestation and conservation ideas, this book seeks a middle ground by exploring the dynamics of these two narratives—in other words, the contestation of Chinese conservationist attitudes. It also has the potential to fill in some gaps in the field of environmental history by (1) providing a rhetorical and intellectual interpretation of why environmental conservation usually does not come to fruition; (2) probing how modern conservationist attitudes were crafted; and (3) showing the literary and cultural impacts of modern Chinese environmental scientists' endeavors.

The tendency to neglect these issues results from a long-standing reluctance on the part of both literary and historical studies to probe *modern Chinese conservationist ideas*. By integrating ecocriticism (literary studies of the environment) with environmental history, this book strives to explore the dynamics at the core of modern Chinese environmental ideas and their impact up to the present day. Writings by Chinese foresters and environmental scientists are sources that have been overlooked by scholars of modern Chinese literature. These writings have circulated throughout the nation and helped to shape the everyday lives of the masses. Conversely, historical studies of the modern Chinese environment tend to overlook the rhetorical strategies and intellectual repercussions of these writings.[19] Whether originating from foresters and scientists or literary writers, these environmental attitudes have consciously or unconsciously shaped environmental actions.

Historical agents in modern Chinese history have used discursive devices to fuel civic action and spark revolutions.[20] Foresters and literary authors spread environmental values through their writings, championing modern perspectives of nature and introducing new environmental values. Their narrative strategies have been both aesthetic and instrumental in shaping public discourse and inciting civic action, meriting in-depth analysis.

To arrive at a full picture of environmental attitudes in modern Chinese society, *Contested Environmentalisms* employs materials from literary works, newspapers, magazines, archives, film and visual culture, scientific writings, and digital media. Nevertheless, this book is less an encyclopedic account than an exploration of the discontinuities, contradictions, and (mal)adaptations of modern Chinese environmental notions. The investigation of the saga of trees and conservationist ideas are used as a lens for probing the history of political mobilization, scientific popularization, and identity formation in modern China.

Modern Chinese Conservationist Ideas in Historical Context: What Is *Modern*?

Largely influenced by the West, modern Chinese environmental ideas have marked a departure from traditional ways of perceiving the natural world. In Chinese history, narratives surrounding trees have been more closely associated with political and moral teachings than with promoting environmental ideas. Dating back to 4000 BCE, the practice of feng shui—literally translated into English as "wind-water"—integrates the layout of natural or built environments with fortune-telling. Based on the notion of harmony among natural elements, feng shui is used to project and increase wealth, happiness, and longevity.[21] For example, trees planted around a building may contribute to the success of a person, a family, or even a corporation inhabiting it.[22] Indeed, in ancient Chinese literature and culture, the depiction of trees had little relevance to environmental conservation per se, even though the benefits of trees were sometimes briefly documented. In the oldest extant collection of Chinese poetry, the *Book of Songs* (*Shijing*), dating from the eleventh to the seventh centuries BCE, there is a poem that uses the image of a woodcutter chopping to criticize rulers who exploit woodcutters' manual labor to lead luxurious lives.[23] Rather than dwell on ecological benefits, the philosopher Mencius (372–289 BCE) uses conservation ideas to provide insight into political governance: "If you don't interfere with the timing of the farmers, there will be more grain than can be eaten. If fine-mesh nets are kept out of the ponds and lakes, there will be more fish and turtles than you can eat. If loggers are regulated in their woodcutting, there will be more timber than can be used."[24] Mencius goes on to define what he assumes to be the proper type of political leadership: "When there is more grain, more fish and turtles than can be eaten, and more timber than can be used, the people will nourish the living and mourn the dead without resentment. Nourishing the living and mourning the dead without resentment is the beginning of the road to true kingship."[25] Renowned poet and politician Liu Zongyuan (773–819 CE) in the Tang dynasty (618–907) similarly invokes trees to convey political thoughts as he argues that one should not attend to a newly planted tree obsessively: if a tree is checked too often and is not offered sufficient freedom to grow, it will perish.[26] Liu's metaphor is meant to show that rulers should grant their people the freedom to flourish independently: if a ruler monitors the ways people sow seeds, plant crops, or raise children, the overmanaged society will fail to flourish. Rather than elaborate on the concrete benefits of trees, Liu uses tree planting as an analogy to illustrate his political ideas.

While trees and forests were largely used as symbolic representatives for moral and political teachings in early- and middle-period China, tree planting began to gain traction in the early modern era, albeit mostly on a regional scale, commercially focused, and largely driven by material interests. According to Ian M. Miller, China began to exercise deliberate afforestation between 1000 and 1600, a period during which a unique system of forest management emerged that was a precursor to modern forestry regimes.[27] Taxing landowners who planted trees in southern China gave rise to a privatized and market-based forestry. Still, tree planting was less a conservation-oriented practice than a new form of agriculture since landowners planted only a few species of trees, aiming at shipbuilding, palace construction, or providing materials for human use. Zhang Meng notes that timber commercialization fueled afforestation around the Yangtze River region in Qing China.[28]

It was probably not until the Qing Empire that the seeds of conservation ideas began to appear in the wake of mounting environmental crises. One poem in *Qing Bell of Poesy* (*Qingshiduo*), written in the late nineteenth century, depicts deforestation as becoming so severe that it wears out both Heaven and Earth.[29] Faced with the rising challenges, the Qing Empire attempted to conserve the natural environment in its borderland. In Manchuria in the period 1750–1850, when ginseng and sable became increasingly scarce, the Qing court conserved them by restricting access to ginseng-producing mountains and banning farming.[30] In regard to forestry, David Bello suggests that the Qing court planted trees to establish its exclusive royal spaces. While such conservation efforts may seem akin to projects being undertaken today, these measures were motivated more by the desire to conserve commodities for the nobility than to protect ecosystems for posterity.[31] It should also be pointed out that Qing rulers' tree-planting practices were undertaken mostly for cultural reasons rather than for their ecological benefits. More importantly, the conservation efforts were confined to certain regions because of a limited scientific understanding of trees' benefits to the natural environment.

At the turn of the twentieth century, Chinese environmentalism began marching into the modern era, developing into what I call *green social engineering*. Four closely linked features characterize modern Chinese environmental notions: a scientific understanding of the natural world; treating environmentalism as a systematic academic discipline; promoting conservationist ideas on a national scale; and, most importantly, cultivating an active, forward-looking citizenry. A new scientific method, offering testable projections regarding the environment, reshaped visions of nature from political allegory to the processes underlying nature's dynamics. Through the work of Chinese foresters, politicians, and intellectuals, trees became the key to resolving multiple urgent issues. For example, these experts saw tree planting as the most efficient and effective means of mitigating natural disasters, especially droughts and floods (see chapter 1). Their belief in science, determination to call for state intervention in the environment, and eagerness to improve China's international position by managing natural resources reveal a modern way of conceiving nature at a time when China was struggling to survive among the global powers. Newly founded academic institutions nationwide promoted rising environmental sciences, such as forestry, ecology, and hydrology, ushering in a scientific and systemic vision of the natural world. These novel epistemological convictions and concerted efforts not only gave rise to new educational institutions to disseminate the benefits of tree planting across China but also, and more importantly, galvanized civic action on a national scale. Tree planting has served as a social-engineering project to mold the Chinese people to become more civilized, disciplined, forward-looking, healthier, and intellectually connected.

Trees, Nationhood, and Subjecthood

In the Western context, Robert Pogue Harrison's pioneering work shows that different historical periods have had distinctive ways of seeing forests. Forests gave rise to isolated cultures in ancient times, led to aristocrats' adoption of hunting as a preferred form of recreation in the medieval period, and became an object to be exploited for economic and social development since the Enlightenment.[32] Beginning in the late eighteenth century, different German groups started to associate the nation's fate with its woodlands, using the "German forest" as a national symbol tailored to their specific social, economic, and political agendas.[33] In American history, forests have been essential to forging the identity of the United States. Two competing metaphors—the forest as howling wilderness and the forest as a holy cathedral—took root in American literature and culture.[34] European settlers in the early seventeenth century took the former view. Roderick Nash's 1967 classic *Wilderness and the American Mind* shows that the harsh wilderness functioned as a foil for and test of Americans' virility and toughness during this period.[35] By the early nineteenth century, influenced by European romanticism, a romantic understanding of nature began to gain traction. Henry David Thoreau, for example, argues that forests serve as cathedrals for transcendental truth.

During the Great Depression of the 1930s, President Franklin D. Roosevelt (1882–1945) established the Civilian Conservation Corps to provide jobs in natural-resource conservation, "putting trees at the center of the fight to save the nation's economy."[36] Three million young men were mobilized to plant trees and to work in forests and parks.[37] Planting trees manifested the American value of hard work as a way out of hardship, contributing to the toughness, resilience, and maturation of youth in an exceptional era in American history. Trees and forests have thus contributed to shaping the American personality and charting US historical paths.[38] So too this has been the case in China.

Tree planting quickly gained impetus during periods of political hardship and social chaos in China. Fulfilling the same role for Americans during the Great Depression, narratives involving trees have spoken to China's desire to secure its survival and prosperity. In the turbulent times of the Republican era, tree planting was regarded as instrumental to the prosperity of Chinese business and industry. Foresters promoted afforestation to remedy shortages of food, clothing, and fuel and to reduce dependence on foreign products. The Yale-educated Ling Daoyang—the pioneering forester in modern Chinese history—claims that importing timber to build railroads paralyzed the economy and impoverished the people.[39] A similar nationalistic logic regarding the benefits of trees extended into the socialist era. In 1956, a time when China was struggling to secure its agricultural production, Mao Zedong launched the Greening the Motherland Campaign, which mobilized youth to plant trees as a way to get them involved in socialist construction. In the 1980s, when uncontained forest fires plagued northeastern China and destroyed at least sixty thousand homes, the widespread destruction of trees forced the authorities to reconsider their reforms and reevaluate their policy of pursuing prosperity.[40]

Tree planting has also shaped the political psychology of the Chinese people. In his 1947 book *The Omnipotence of Forests* (*Senlin wannenglun*), Hao Jingsheng (1903–55), one of the leading foresters in Republican China, argues that because forests embody a nation's natural resources in its soil, advocating for afforestation nurtures citizens' love of their country.[41] Hao deploys emotional appeals, including patriotic sentiments, a sense of crisis, and a promising vision, to advance conservationist ideas as well as instill nationalistic sentiments in the citizens. The emotionally invested forest served as one prong of a widespread appeal to sentimentality deployed in shaping public opinions and civic actions during the first half of the twentieth century, contributing to what Haiyan Lee calls "universalizing norms of enlightenment humanism and nationalism," or what Eugenia Lean describes as "collective emotionalism."[42]

Paradoxically, however, tree planting not only unites but also divides the Chinese people. When tree planting and its associated environmental conservation ideas became internalized in the Chinese mind, those who espoused these ideas became subjects of the Chinese nation-state. For a young recruit in a 1950s short story, identifying enemies who may harm the forest indicates that he is on the threshold of becoming a mature Chinese soldier (see chapter 4). For ethnic Mongolians portrayed in ecocinema, resistance to tree planting suggests refusal to be controlled by Han-dominated Communist China rather than a simple rejection of environmental ideas (see chapter 3).

The circulation of conservationist ideas also reveals China's ambition to gain and secure its position in a changing world.[43] During the isolation of the Cold War era, China sent documentaries on conservation to world forestry conferences, striving to present a positive image of the country. In response to the global outcry over China's greenhouse-gas emissions, its biggest e-commerce company, Alibaba, launched an initiative designed to promote tree planting: each time cellphone users select specific eco-friendly businesses or activities, Alibaba makes a small contribution toward a desert-reforestation project (see the coda). In 2020, while the United States was exiting the Paris climate agreement under the Donald Trump administration, China claimed that its afforestation projects dramatically contributed to curbing global carbon dioxide emissions and moving to net carbon neutrality by 2060. In response to critiques from other nations, China has characterized its afforestation projects as a symbol of its endeavors to promote the protection of nature for the entire world.

Nevertheless, this book suggests the anthropocentric nature of modern Chinese conservation efforts, which prioritize state interests over ecological restoration. It particularly focuses on tree-planting initiatives, framing them within Bryan Norton's concept of "weak anthropocentrism"—a perspective that, while human-centric, recognizes the importance of environmental conservation for humanity's long-term benefit.[44] The book raises concerns about the ecological consequences of such practices, especially in desert areas. It suggests that planting trees in deserts—a topic explored in detail in chapter 3—might not be as environmentally friendly as it appears.[45] Planting trees in the desert can also be regarded as an invasive process to disrupt and harm the original environment, which has its own unique desert ecosystem. The state's interests often overshadow the needs of the natural world, indicating that despite their adaptability, these strategies have inherent limitations. In essence, the book argues that conservation in modern China, at its core, remains predominantly focused on human interests rather than on the preservation of nature.

Historicizing Chinese Authoritarian Environmentalism

Recent scholarship has begun to acknowledge the environmental initiatives promoted by authoritarian regimes.[46] This line of scholarship, mostly rooted in political science, highlights the interplay between environmentalism and authoritarianism in contemporary China.[47] While these works are commendable for their exploration of the oppressive, multivalent, and performative nature of Chinese environmental politics, they open up space to understand how authoritarian environmentalism advances to its current status. This book historicizes authoritarian environmentalism in modern China to present a comparative perspective on global environmental governance in shifting and diverse cultural contexts.[48] *Contested Environmentalisms* demonstrates the intertwining of several key factors—transnational science, mass mobilization, local knowledge, economic interests, and geopolitical climate—and how they collectively fortify the resilience of authoritarian China. Each component plays a vital role in bolstering the nation's authoritative structure, showcasing the dynamics that contribute to its strength and stability.[49]

One strength of Chinese authoritarian regimes has been their ability to assimilate the knowledge produced by democratic environmentalism in Western countries and even to allow contested voices to coexist. This has contributed to China's phenomenal rise in recent decades. Compared with their authoritarian counterparts, especially those in the former Soviet Union, Chinese leaders have taken a more flexible approach to adjusting their economic and social policies according to changing political climates. While maintaining their control of Chinese society, they opened up the economy to foreign investments and encouraged international cooperation in many fields. As we will see, Western science and conservationist ideas have circulated and played a pivotal role in Chinese policymaking and public rhetoric—even during the Mao era.

Far from being incompatible, authoritarian and democratic environmentalism are, I argue, similar and complementary. While authoritarian environmentalism is usually efficient but is imposed on the populace by a powerful central state, environmentalism in liberal democratic representative regimes is highly affected by lobbying from business interests and responds slowly to public concerns.[50] Nevertheless, each can draw on the other's strengths: authoritarian regimes may adopt elements of free market economies to popularize conservationist ideas, while democratic environmentalism can evoke collective efforts to advance a conservationist agenda. Viewed through the lens of history, it is evident that during a crisis, the prompt and cohesive action characteristic of an authoritarian regime often yields greater efficiency. The aforementioned Civilian Conservation Corps spearheaded by President Roosevelt integrated widespread public participation with a relatively strong federal power in a time of crisis. In this way, the United States integrated collective efforts mostly characteristic of authoritarian regimes to centralize and expedite environmental practices at one of the most critical junctures in US history.[51] In less pressing situations, democratic environmentalism usually thrives as its longevity and influence are rooted in core principles: fostering inclusive public engagement in environmental policymaking, promoting transparency, and maintaining strict accountability for results. Democratic environmentalism often offers incremental improvements, fostering deeper heartfelt trust and faith in environmental practices among the participants. Such an enduring impact is mirrored in initiatives like China's *Ant Forest* (*Mayisenlin*), to be discussed in the coda, which displays the potential for wide political participation to resonate deeply and lastingly with the Chinese populace. Thus, the dichotomy between authoritarian and democratic environmentalism is by no means absolute.[52] Historical contexts necessitate that each polity incorporates essential characteristics from its counterpart, highlighting that these two approaches, while distinct, are not diametrically opposed; instead, they offer diverse routes that ultimately converge upon the common goal of a sustainable future.

This book also suggests that largely regardless of political systems, different knowledge systems relating to the natural environment make certain types of environmental politics possible. Distinctive knowledge systems are generated by people of diverse backgrounds, including environmental scientists trained in various academic disciplines, writers influenced by the traditions either of a foreign country or their own, and grassroots organizers who have faith either in universalist rhetoric or in Indigenous or personal views. These different sources of environmental ideas—which are deeply intertwined with academic background, class, gender, and ethnicity—are shared features of both authoritarian and democratic bureaucracies.

In both forms of society, too, unaccountable bureaucratic power plays a pivotal role in environmental problems. Kate Brown's *Plutopia: Nuclear Families, Atomic Cities, and the Great Soviet and American Plutonium Disasters* demonstrates how bureaucratic systems in the Soviet Union and the United States similarly neglected public-health concerns while advancing their nuclear-weapons projects.[53] Although it focuses on China, this book shows how Chinese environmental discourse can be deployed to challenge bureaucratic systems and address social injustice more generally.

Despite the possibility of negotiating with democratic environmentalism, Chinese authoritarian environmentalism has intrinsic limitations. Though it is rapid in making and implementing its policies, the environmental attitudes it promotes are usually short-lived. For example, founded in 1978, the Three-North Shelter Forest program, a government centralized effort aimed at protecting soil from erosion in northern China, did not drive home the message of environmentalism.[54] And even though it was promoted by top Chinese leaders, Arbor Day ultimately fell into oblivion. More recently, Zhang Xinqi, the "tree-planting mayor" of China's coastal city Qingdao, waged a massive afforestation campaign. In 2012, he budgeted 4 billion yuan (around 6 million US dollars) to plant trees. Because of insufficient political participation in making this decision, however, the campaign received a frosty reception from local citizens, who complained that the tree planting disturbed their daily lives. After Zhang was removed from his position owing to corruption allegations, the campaign was suspended.[55]

All of this points to a basic problem of authoritarian environmentalism in China and elsewhere: a centralized polity is in constant tension with the need to popularize environmental ideas among the masses. As these regimes find other initiatives to focus on, the environmental endeavor often loses momentum once a specific campaign is over. Because the dissemination of environmental ideas is largely determined by a few supreme leaders who may step down or change their minds, an authoritarian regime cannot offer a sustained and coherent solution to environmental issues in the long term.

In China, this problem of frequent turnover results from the campaign-style environmental governance explored in this book (chapters 1 and chapter 2). A centralized campaign effort to promote environmental conservation is inherently at odds with the urge to bring environmental ideas to the masses spread across China's vast territory. A short-term campaign to cultivate environmental consciousness clashes with the long and incremental work of putting environmental ideas into practice in everyday life.[56] Once the campaign is over, conservation practices come to a halt. Even though indoctrination in environmental notions may continue, its influence pales in comparison to a cause to which individuals are personally committed. The campaign-style strategy has consumed too much of the public's energy, rendering everyday education in conservationist ideas less effective.

In addition to these problems originating at the top, authoritarian environmentalism does not encourage bottom-up political participation, making it hard to mobilize the populace. The environmental notions circulating in the country are always those promoted by the central government, which reveals that the putative universality of Chinese conservationist ideas is a myth. Voices of environmental conservation that weaken or even attack the government's social control are suppressed.[57] The manufactured universality of Chinese environmentalism engenders a paradoxical result: the more authorities circulate their approved environmental ideas, the less likely those ideas are to create heartfelt and enduring effects on the general public. As a result, modern China's environmentalisms, because of their authoritarian DNA, are constantly emerging and disappearing. The tension between centralized policymaking and a decentralized imperative to disseminate environmental ideas among the masses not only shows the limits of popularizing conservationist attitudes but also demonstrates how environmentalism operates as a potent and enduring historical force.

The Contour of Contested Environmentalisms: Chapter Summary

Arranged in chronological order and drawing on literature (including science fiction for children, reportage literature, and diaspora writings), films, newspapers, archives, posters, scientific journals, and digital media, the chapters of this book unveil various aspects of environmentalism in Chinese culture in the period 1912–2020. While each chapter has a specific focus, the chapters also engage with each other to show the contestation of modern Chinese environmentalism in a transnational context.

Chapter 1 traces the rise and fall of modern environmentalism in the Republican era (1912–49) by focusing on Arbor Day. Using scientific documents, news reports, journals, and archives, this chapter demonstrates the tension between tree planting as a practical activity and as a symbolic gesture. Advancing modern forestry as a new academic discipline, Western-educated Chinese foresters strove to promote tree planting to boost the national welfare, culminating in the first observation of Arbor Day in 1916. Then, the chapter discusses how Arbor Day became an increasingly symbolic gesture, focusing mainly on celebrating Sun Yat-sen: while both Chiang Kai-shek and Wang Jingwei deployed tree planting to show their political legitimacy as heirs of Sun, tree planting's ecological benefits were increasingly overlooked. The final section traces conservationist rhetoric of the Communist regime. Though the Communists did not celebrate Arbor Day during the Republican era, their tree-planting initiatives were indebted to the environmental endeavors of foresters and politicians of that time. The story of the tree-planting efforts by the Nationalists and the Communists offers an entry point into examining the generation of modern Chinese conservationist ideas and uncovers the roots of the environmental ideas of the Communist government that are discussed in later chapters. This chapter also shows how, from a global perspective, environmental ideas are always fluid and adaptable. The dissemination of belief in the benefits of trees in Republican China was a transnational project of domesticating Western forestry science, and more importantly, it embodied a web of negotiation among disparate Chinese political and social factions.

Chapters 2 through 4 explore scientific texts, literary works, and films that served as key generators of modern Chinese environmental ideas in Mao-era China. Chapter 2 examines the rhetoric of foresters' writings and speeches as well as socialist propaganda during the early socialist era (1949–61). It first explores the speeches and writings of Liang Xi, probably the most important forester and forestry official in early socialist China. He strategically promoted tree planting in the early 1950s, clashing with hydrologists who also aspired to solve China's environmental challenges. Despite the fact that both hydrologists and foresters are environmental scientists, they have worked from drastically different visions of nature. This chapter then examines the rhetoric of the Greening the Motherland Campaign launched by Mao in 1956. During this initiative, Mao pushed the Yellow River tree-planting initiative to a national scale thanks primarily to the dedicated persuasive endeavors of the foresters. This nationwide initiative, aligned with the new government's nation-building activities, necessitated the imparting of tree-planting expertise on the part of foresters to a wide spectrum of individuals at the grassroots level while also strategically incorporating it into the socialist revolutionary and global environmental discourse.

While chapters 1 and 2 examine how foresters' rhetoric gradually became an integral part of Chinese state policymaking, chapter 3 employs films in the Mao era to explore how this afforestation rhetoric came to the Chinese borderlands and the reactions of ethnic groups to it. In short, the Greening the Motherland Campaign in the borderlands caused the Han Chinese conception of nature to clash with that of ethnic minorities. I argue that undertaking tree planting as a universally progressive crusade without considering local environmental traditions is inherently regressive because it overlooks Indigenous perspectives on the natural world. The first section of the chapter explores the scientific discourse concerning the conquest of the desert promoted by the Communist government. The treatises depicted tree planting as a panacea, a beneficial method that could be widely adopted. The second section examines a feature film, *Spring in the Desert* (*Shamo de chuntian*; 1975), which illustrates how resistance from the local ethnic population emerges when universalist values are at odds with local ones held by ethnic groups. I contend that tree planting in the desert reflects the conflict between the settled Han Chinese lifestyle and the nomadic way of life embraced by the Mongolians. The third section probes a documentary film, *The Army's Reclamation and Battle Song* (*Junken zhange*; 1965), which attempts to justify the state's social dominance in Xinjiang under the guise of environmental conservation.

Chapter 4 links the preceding and following chapters by exploring how conservationist rhetoric in the Mao era began to enter the literary imagination and expand in public rhetoric through a process of demilitarization. This chapter shows that the military mindset, in addition to the scientific rhetoric explored in earlier chapters, has also motivated the emergence of conservationist ideas. Killing enemies has not been merely aimed at eradicating imminent threats; killing enemies hidden in the forests has confirmed the Communists' right to the land, which allowed them to conserve natural resources in later decades. Studying two Mao-era works, "Songs of the Forest" ("Senlin zhige"; 1955) and *Tracks in the Snowy Forest* (*Linhaixueyuan*; 1957), this chapter first probes how the Communists have negotiated romantic and militarized understandings of nature as reflected in the stories of young soldiers gradually coming of age. The works show how socialist China aspired to nurture a purified forest and regarded this purified environment as a barometer of socialism. Finally, the chapter analyzes two science fiction short stories that express conservationist concerns after human enemies have fallen into oblivion and nonhuman enemies are under control. While the narrative mode focusing on identifying the Nationalist enemies begins to wind down and becomes less confrontational, these sci-fi works signal a more explicit desire to conserve the natural world—a trend that becomes more evident in literary works of the 1980s, examined in the subsequent chapter.

Chapter 5 argues that despite the rise of conservationist rhetoric in the 1980s, two opposing trends—revisiting Chinese native traditions (green nativism) and celebrating Western environmental conservation achievements (green Occidentalism)—did not succeed in conveying the message of environmental protection to the public. The first section explores Xu Gang's nonfiction work *Woodcutter, Wake Up!* (*Famuzhe, xinglai!*; 1988), which celebrates Western science and environmental achievements as a way to solve Chinese ecological problems. The following section examines Ah Cheng's novella *The King of Trees* (*Shuwang*; 1985), which reveals how China's own folk traditions can become a solution to the country's ecological woes.[58] The chapter explores how, because of their distinctive environmental thoughts, these two voices were not able to merge effectively to raise citizens' awareness of environmental degradation. A second argument made in this chapter is that the rise of what I call *Chinese conservation literature* is both the *beginning* of a new era of environmental consciousness and the *consequence* of the century-long effort, investigated in the previous chapters, to instill environmental awareness among the public. By examining textbook materials, scientific journals, propaganda works, and interviews, the chapter also delves into how the scientific discourse of trees and forests penetrates modern Chinese literature.

The final chapter, the coda, turns to the most recent tree-planting initiative, which uses digital media. In 2016, the e-commerce giant Alibaba launched its *Ant Forest* app to encourage its customers to plant trees via their cellphones. Thus, just over one hundred years after the Republican era celebrated China's first Arbor Day, one of the country's most influential companies initiated a virtual Arbor Day. Unlike the one in 1916, which took the form of a one-day ceremony, this tree-planting initiative has encouraged environmental awareness by continually evoking the socialist rhetoric of bottom-up participation by the public. I argue that in deploying this rhetoric, *Ant Forest* strikes a balance between socialist mass mobilization and capitalist moneymaking, fosters active citizenship by instilling warm and compassionate feelings in its users, and diminishes the necessity for environmental knowledge in advancing tree planting. Finally, the chapter recaps the arguments of the book to claim that contestation operates as a source of strength rather than weakness for Chinese environmentalism.

Exploring key moments in Chinese environmental ideas from 1912 to 2020, *Contested Environmentalisms* shows the continuity, ruptures, and challenges of disseminating conservation notions. In this history, trees and forests serve as a key to resolving China's environmental problems and forging a science-oriented, emotionally bonded, and forward-looking spirit in this rising global power.

ONE

Arbor Day

The Rise and Fall of China's Environmentalism in the Republican Era (1912–49)

Introduction

"We plant trees and turn the nation green," a child in Nanjing loudly proclaimed during Arbor Day in 2013.[1] The Communist Chinese government had not informed the child that March 12, the date on which he celebrates Arbor Day, also marks the death of Sun Yat-sen (Sun Zhongshan; 1866–1925), the founding father of Republican China (1912–49). This coincidence points to a little noticed but crucial link between the Republican and the current Communist governments regarding tree planting, indicating a century-long saga of instilling conservationist ideas in the Chinese general public. This chapter considers one of the fundamental manifestations of modern China's environmentalism: Arbor Day and its variations. Arbor Day reflects a crucial tension in Chinese environmentalism between nature and culture. On the one hand, it is an expression of the desire to belong to nature; by planting trees, the Chinese show their eagerness to naturalize their surroundings. On the other hand, Arbor Day reflects the need for individuals to be part of the culture; through group participation, individuals establish and affirm their social identities. This inherent tension manifested itself in a convoluted and even ironic manner in the Republican era, when tree-planting exercises became increasingly symbolic.

An examination of the founding of Arbor Day and its aftermath is essential to understanding the generation and circulation of Chinese conservationist ideas as well as their roles in forging Chinese modernity in the early twentieth century. While scholars have explored Chinese modernity from multiple perspectives—such as linguistic, hygienic, and economic—an environmental aspect of modernity, especially its inherent tensions, illuminates not only modern Chinese intellectuals' sincere and determined pursuit of lifting tens of millions of people out of poverty but also the contingent and adaptive trajectory of modern Chinese environmental attitudes in the early twentieth century.[2]

This chapter first traces the endeavors of Chinese foresters and Sun Yat-sen to promote modern environmentalism. It situates China's environmentalism in a global context by linking its forestry to happenings in the United States and other Western nations. It shows that Western forestry science was introduced to China during a national crisis. It then traces the evolution of environmental ideas in modern China in two conflicting camps: the Nationalists and the Communists. While both are indebted to Sun, they inherited his legacy in different ways, which accounts for their disparate attitudes toward the natural world.

This chapter highlights that Chinese environmentalisms are not only plural but also the subject of intense contestation. At the beginning of the twentieth century, Western-educated foresters were actively replacing the old modes of perceiving nature handed down from ancient China, leading to the establishment of Arbor Day. Sun Yat-sen's efforts to promote both tree planting and tree cutting reflect his vision for national renewal and survival during times of crisis. After his death, the political rivals Chiang Kai-shek (Jiang Jieshi; 1887–1975) and Wang Jingwei (1883–1944) both celebrated Arbor Day in an attempt to establish themselves as Sun's legitimate successors. The Nationalists' symbolic observance of Arbor Day stands in contrast to the Communists' emphasis on actual tree planting. These historical agents' endeavors not only reveal the dynamic contestation of modern Chinese conservationist ideas but also uncover the roots of environmentalism in the Communist government in later decades.

The Rise of Modern Chinese Environmentalisms: A New Academic Discipline and the Establishment of Arbor Day

The establishment of Arbor Day was closely associated with China's century-long engagement with the West, when late Qing intellectuals, Western missionaries, and Chinese foresters endeavored to introduce Western ideas of forestry into China. They advanced a novel and scientific view of nature in which trees became pillars of environmental management.

Western conservationist initiatives began to take root in China beginning in the late Qing period (1840–1911). Proposals to plant trees gained traction when late Qing scholars sought to introduce Western forestry to help solve China's social and ecological ills. Though ancient China had unsystematic conservationist ideas, as I have suggested in the introduction, they neither sponsored forestry as an academic discipline nor elevated it to a national-scale development priority until the late Qing. In 1878, *Globe Magazine* (*Wanguo gongbao*), endorsed by Western missionaries, published an article titled "Tree-Planting School" ("Zhishu xueshu"). The article expresses regret that China did not have the private institutions to promote forestry, as was the case in Europe.[3] Unlike the United States and the British Empire, which had deforested much of their land, the article notes, the Germans advanced the benefits of trees, which included moderating the summer heat, avoiding floods and drought, and safeguarding agricultural production. Germany had set up forestry as a discipline, producing more than a hundred books on tree planting per year.[4] The article suggests that China should follow Germany's lead in promoting tree planting for the benefit of future generations. Prominent late Qing intellectuals found this argument persuasive: reformers such as Zhang Zhidong (1837–1909) and Kang Youwei (1858–1927) were under the sway of Western ecological ideas. Zhang submitted a written statement to the Qing emperor in 1901 advocating a bonus for officials who plant trees in barren regions, and Kang asked the Qing government to found an institution in charge of forestry.[5]

While these late Qing reformers articulated the need to develop forestry, in practice, it was the US-educated foresters who systematically introduced modern forestry to China.[6] Receiving his master's degree in forestry from the University of Michigan in 1911, forester Han An (1883–1961) argued that forestry is a symbol of a nation's civilization and development.[7] Pointing out that Western forestry as a scientific academic discipline had helped to provide sufficient timber for Western countries' national development, he suggested that China could follow a similar path.[8]

Another pioneering Chinese forester, Ling Daoyang (1888–1993), was heavily influenced by an iconic figure in US forestry, Gifford Pinchot. After graduating from Yale University, studying forestry in France, and becoming the head of the Division of Forestry of the US Department of Agriculture, Pinchot spearheaded the founding of the Yale Forest School at his alma mater in 1900, setting the tone of US environmental conservation as "wise use": though stressing the need to protect the natural environment, he believed that public wilderness areas could be used as a source of income for the country if the resources were handled wisely. The courses at Yale were influenced by Pinchot's environmental attitudes, which integrated both the utilization and conservation of nature. In particular, the curriculum offered "a preparation not merely for practical work in the woods, but also for the broad work of forest organization in Government and State service, for handling large tracts of forest land, for consultation work for railroads, lumbermen, water companies . . . , for teaching, and for scientific research."[9] The courses covered timber production as well as forest conservation, including silviculture, forest products, forest management, wood preservation, and the history of forestry.[10] Pinchot and his peers hoped that the careful management of the nation's resources by trained professionals would lead to prosperity for the American people for generations.[11] As the first US institution to have a forestry school, Yale boasted many illustrious graduates who went on to become the leading foresters of the world, including Henry S. Graves. Ling Daoyang took courses at Yale on forest utilization, including three courses on lumbering and five on forest products, and on forest conservation, including three courses on silviculture (see figure 1.1).[12] Earning his master's degree in forestry in 1914, he returned to China to popularize forestry as a full-fledged academic discipline. He emphasized the conservation or protective aspects of forestry—tree planting in particular—once he took office as a specialist in the Ministry of Agriculture and Commerce and chairman of the University of Nanjing's Forestry Department.

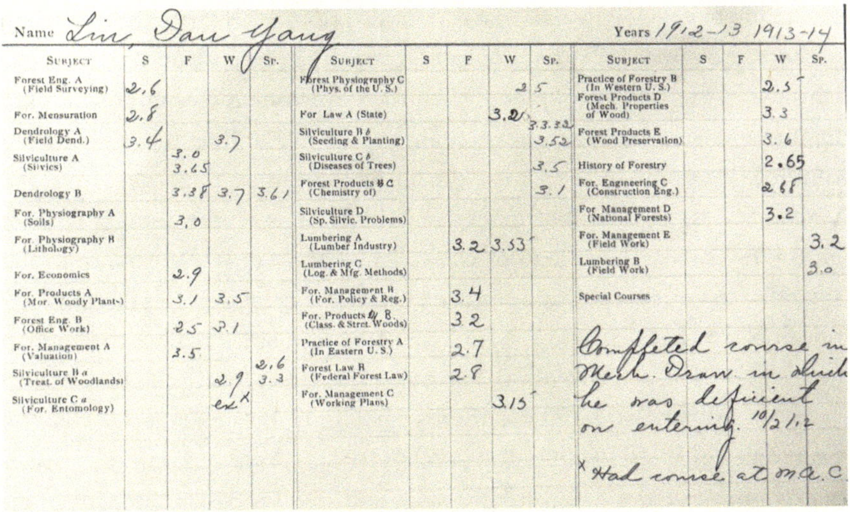

Name Lin, Dau Yang   Years 1912–13 1913–14

Subject	S	F	W	Sp.
Forest Eng. A (Field Surveying)	2.6			
For. Mensuration	2.8			
Dendrology A (Field Dend.)	3.4		3.7	
Silviculture A (Silvics)		3.0 3.65		
Dendrology B		3.38	3.7	3.61
For. Physiography A (Soils)		3.0		
For. Physiography B (Lithology)				
For. Economics		2.9		
For. Products A (Mor. Woody Plants)		3.1	3.5	
Forest Eng. B (Office Work)		2.5	3.1	
For. Management A (Valuation)		3.5		
Silviculture B *a* (Treat. of Woodlands)			2.9	2.6 3.3
Silviculture C *a* (For. Entomology)			ex×	

Subject	S	F	W	Sp.
Forest Physiography C (Phys. of the U. S.)				2 5
For Law A (State)			3.2	
Silviculture B *b* (Seeding & Planting)				3.3.32 3.52
Silviculture C *b* (Diseases of Trees)				3.5
Forest Products B C (Chemistry of)				3.1
Silviculture D (Sp. Silvic. Problems)				
Lumbering A (Lumber Industry)		3.2	3.53	
Lumbering C (Log. & Mfg. Methods)				
For. Management B (For. Policy & Reg.)		3.4		
For. Products B (Class. & Strct. Woods)		3.2		
Practice of Forestry A (In Eastern U. S.)		2.7		
Forest Law B (Federal Forest Law)		2.8		
For. Management C (Working Plans)			3.15	

Subject	S	F	W	Sp.
Practice of Forestry B (In Western U. S.)			2.5	
Forest Products D (Mech. Properties of Wood)			3.3	
Forest Products E (Wood Preservation)			3.6	
History of Forestry			2.65	
For. Engineering C (Construction Eng.)			2.65	
For. Management D (National Forests)			3.2	
For. Management E (Field Work)				3.2
Lumbering B (Field Work)				3.0
Special Courses				

Completed course in Mech. Draw. in which he was deficient on entering. 10/21/12

× Had course at M.A.C.

FIGURE 1.1 **Yale's academic record of Ling Daoyang, the pioneering modern Chinese forester. Source: Alumni Records Office, series 11, "Forestry," box 1201, courtesy of Beinecke Rare Book and Manuscript Library, Yale University.**

In addition to Chinese foresters studying abroad, Western foresters visited China, where they contributed not only to modern Chinese forestry but also to social reform.[13] American missionary Joseph Bailie, for example, enthusiastically promoted forestry education in China. He argued that forestry could prevent famine and help to relieve the burdens of Western medical missionaries working in China.[14] Bailie founded the Colonization Association in Nanjing, which pursued famine relief for the poor by asking peasants to plant trees.[15] He contended that sustained agriculture could transform bandits into law-abiding citizens and that forestry was an essential step to fulfilling this task.[16]

Bailie's request for the United States to sponsor China's first forestry school in Nanjing in the 1910s was tinged with a desire to increase US presence in China during the First World War (1914–18). In an article published in *American Forestry*, he viewed forestry from the perspective of the Sino-US relationship: "The Chinese nation is unconquerable. It swallows up its conquerors. . . . America has the confidence of every Chinese of intelligence both in the United States and in China. Any approaches now made to help the nation to develop its national resources and then alleviate the condition of the worthy poor will be appreciated and backed by cooperation from the Chinese."[17] He was encouraged when the government closed its school in Beijing and sent its best students to his school in Nanjing. In order to offer continued support to the Chinese, he called for more donations from the United States, arguing that "America as a nation is on the straight road to becoming that bosom friend of China and a grant now in response to the appeal of the minister of agriculture would do a great deal to seal that relationship."[18] When China was engaged in military conflicts, forestry came to be seen as representative of the US presence there. During the First World War, when Japan increased its influence in China, Bailie suggested that the United States should counter by making an economic and political investment in the country.[19] Forestry, he said, was an essential component in accomplishing this goal.[20]

These foresters from China and the West contributed to the establishment of Arbor Day. Ling Daoyang, Han An, and Joseph Bailie wrote to Zhou Ziqi (1869–1923), the minister of the Department of Agriculture and Forestry, arguing that afforestation would be instrumental in advancing China's future.[21] In a document requesting that President Yuan Shikai (1859–1960) approve Arbor Day, Zhou argued that trees contribute to people's livelihoods since they consolidate levees and prevent sandstorms and floods.[22] The document suggested that the president should select a specific place in the capital to plant trees, thereby setting an example for the nation. In 1914, the Republican government under Yuan passed the Chinese Forestry Law (Zhongguo senlin fa), which stipulated the need to expand the protective function of trees.[23] On July 31 of the following year, Yuan issued an executive order designating the Qingming (Tomb-Sweeping Day) festival as Arbor Day. On April 6, 1916, the first Arbor Day celebration saw the absence of President Yuan, who was battling political turmoil and deteriorating health. His representative Wang Da (1881–1946) was sent to a barren land, Xishan, west of Beijing, to attend the event in his stead.[24] The celebration was a highly ceremonial affair, with attendees scaling the mountain and bowing to the national flag before the tree planting. Then, the secretary of agriculture and commerce, his subordinates, and other officials from Beijing planted trees as ordered, making the process a manifestation of political hierarchy. As the Arbor Day event came to a close, photographs were taken to serve as a reminder of the lasting impact of conservation efforts. At the same time, they highlighted the symbolic implications of the ceremony and the crucial role of political order even in the absence of the highest authority.

After Yuan's death during the political violence, the new president, Li Yuanhong (1864–1928), further invested Arbor Day with ritualized connotations. On April 6, 1917, he came to the Heaven Temple (Tiantan) in Beijing to plant trees.[25] The celebration at the Heaven Temple, a location that is traditionally used to praise Heaven and pray for rain during important occasions, imbued Arbor Day with a religious significance that would render it even more symbolic in the near future.

After the establishment of Arbor Day, Chinese foresters continued to advance afforestation. In 1916, Ling Daoyang published his *Main Ideas of Forestry* (*Senlinxue dayi*), which substantiates the benefits of tree planting and aspires to set up forestry as an independent academic discipline. He divided the benefits into two categories: direct and indirect. The direct benefits include timber and other wood products, which attract financial investment and provide job opportunities. Ling was heavily influenced by Pinchot and Graves, who tended to regard forestry as a profession that managed natural resources for economic purposes. Unlike modern-day ideas, which prioritize the ecological benefits of trees, Ling classified those as "indirect," arguing that trees control temperature and moisture. He noted that in Germany, summer temperatures in the forests are seven degrees Fahrenheit lower than those outside the wooded areas.[26] He further argued that forests increase rainfall: for example, in Nancy, France, rainfall was increased by 24 percent thanks to the forest. In addition, forests control soil erosion: tree roots absorb water, and their leaves reduce the impact of the drops on the ground. Ling claimed that with these environmental ideas in mind, France, Switzerland, Hungary, and Austria all passed forestry laws. He also cited studies by his teacher Graves suggesting that forestry contributes to water conservation. He argued that China, a predominantly agricultural country, could boost the production of its farms if a sufficient amount of water was preserved by trees.

Ling's scientific elaboration of the benefits of trees in *Main Ideas of For-*

estry inaugurated a new vision of the natural world, one that characterizes modern Chinese forestry. In contrast to the traditional Chinese understanding of nature—based on local experience and usually used as a political metaphor—modern forestry offers a scientific model that can be applied anywhere.[27] As examined in the introduction, the Confucian philosopher Mencius provides an example of the traditional view. Mencius states that if trees are harvested too quickly, they will not flourish. He does not go deeper into the rationale for this advice but shifts the focus to political theory: just as trees need to be cultivated, so do human hearts. Drawing on early historical documents, the Tang dynasty poet Yuan Zhen (779–831 CE) contends that tree planting could prevent flooding and urges future generations to take up this strategy.[28] Like Mencius, Yuan does not delve into the underlying reasons for the protective function of trees. It was not until the turn of the twentieth century that foresters began to systematically offer scientific explanations of the benefits of afforestation. Replacing locally based and experience-dominated knowledge regarding tree planting, foresters in Republican China developed a universalistic perception of the natural environment that could be applied anywhere in the world, suggesting that Western science might provide an effective means of solving China's problems. Furthermore, as Ling observed, forestry science makes the natural world predictable. Throughout this discussion, Ling employed Western forestry science in an attempt to convince Chinese readers of the effectiveness of modern forestry as an academic discipline.

Ling also argued that forests have curative potential, nurturing physical and mental health. He explained that the air and water in the forest are clean, which helps prevent disease. In Germany, he noted, areas closer to the forest have more oxygen, which contributes to a more pleasant and healthy atmosphere for the citizens. In the United States, according to Ling, wooded city parks encourage the citizens to enjoy the scenery and recharge themselves.[29] He argued that if people do not have forests for relaxation, they will either stay at home or spend their time gambling or engaging in other wasteful pursuits.[30] If immersed in forests, they will cultivate a more civilized, healthy, and forward-looking spirit. Trees can thus become integral to Chinese development by contributing to greater civic-mindedness and public participation.

In his more easily digestible pamphlet *Forest Outline* (*Senlin yaolan*) of 1918, Ling Daoyang articulated the urgency of developing forestry from the perspective of national development. He noted that tree planting had been endorsed by the governments of many countries and said that China should follow suit. He asserted that China was far behind nations such as Germany, the United States, India, Russia, France, Switzerland, and Japan in percentage of forest coverage and should act immediately to catch up. In order to convince his readers, he presented a diagram depicting the gap and illustrating the direct (beneficial to papermaking, medicine, and garment production) and indirect (adjusting moisture, preventing floods, and hosting animals) values of planting trees (see figure 1.2).[31] The diagram conveys

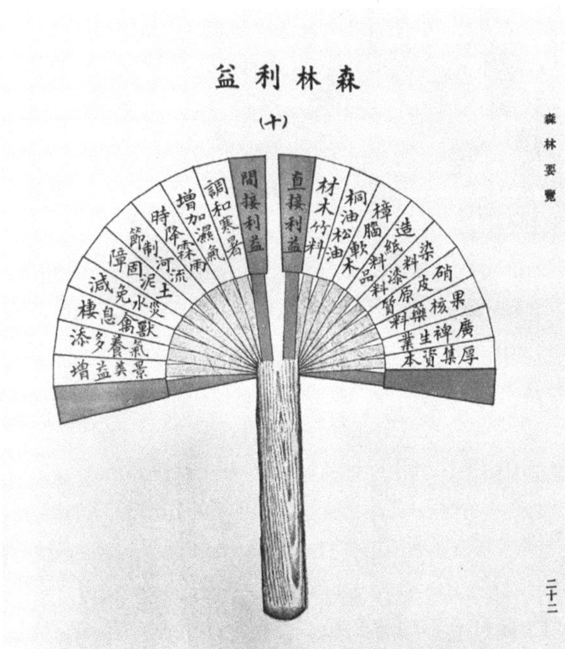

FIGURE 1.2 **Illustration from *Forest Outline* (*Senlin yaolan*), in which Ling Daoyang extolled the benefits of forestry to lay audiences. The left side of the fan lists the indirect benefits of trees, including adjusting temperature, preventing floods, and increasing moisture in the atmosphere. The right side shows the direct benefits, such as providing timber, making paper, and increasing capital. Source: Ling Daoyang, *Senlin yaolan*, 22.**

literal as well as metaphoric meanings. If the trees germinate and grow, the ecological and economic benefits shown under the canopy in the picture will enlarge accordingly.

Rapid modern forestry development led to trees' embodiment of contentious geopolitics in East Asia. In northeastern China, timber became the battleground between the Japanese and Russian Empires. Russia had been exploiting forest resources in the region since the late Qing. Worried about Japanese reactions, Russia was careful about extending its influence southward. Even so, Russian efforts to use timber to build railways and electric poles in the Yalu River Valley unsettled the Japanese, who were eager to increase their own impact on the area. As early as 1898, Japan had begun to survey forestry resources in northeastern China, later using the survey to increase its production of timber to build railways. The Japanese government was upset by the Russian timber-based businesses, which partially militarized the region: for the Japanese, Russia was not merely promoting its forestry management in northeastern China but was also expanding its empire. Japan launched a war against Russia in 1904 and claimed victory in 1905.[32] The victory further stimulated Japan's appetite for forest resources in China. In 1908, the Japanese sponsored the Yalu River Logging Company (Yalujiang caimu gongsi) to exploit those resources. Japanese foresters, like their Chinese counterparts, had acquired modern forestry from the West. The new knowledge increased the Japanese desire for timber and accelerated the utilization of forests in northeastern China. Within thirty years, forest resources were severely depleted.[33]

The competition for Chinese forest resources from the Japanese and Russian Empires inspired Ling to found the China Forestry Association (Zhonghua senlin hui) in 1917. The organization has endeavored to convince policymakers that tree planting is fundamental to making China prosperous.[34] Ling also assisted Sun Yat-sen in creating his "three principles of the people" (*sanminzhuyi*)—nationalism, democracy, and the livelihood of the people—and his "plan for building the nation" (*jianguo fanglüe*), both of which feature forestry.[35]

Tree Planting to Prevent a Revolution: Sun Yat-sen and His Green Cultural Capital

Thanks to the efforts of foresters, Sun Yat-sen came to consider forestry an essential and even fundamental component of China's nation building. In the lengthy and pivotal political manifesto "Three Principles of the People" ("Sanminzhuyi"), published in 1924, Sun articulated the linkage between forestry and people's livelihoods.[36] He argued that forestry can improve living conditions: by storing water and keeping rain from hitting the ground, trees are crucial for preventing floods—a perennial difficulty that has impoverished the Chinese people and depleted the national budget. He further pointed to the danger of social unrest if there are no trees to stop droughts and the consequent famines:

> There were droughts in Russia every two or three years after the revolution. Owing to that drought, a lot of people starved to death. The Russian Revolution almost failed, and we could see that drought was perilous. In the past, the drought could not be avoided. Currently, thanks to science, any kind of natural disaster can be relieved. However, the way to avoid drought should be framed with a comprehensive national plan. What is the solution? The fundamental solution is to plant trees. Once trees are planted, water in the air can be adjusted. It will often rain, reducing drought as a result. . . . So the fundamental methods we have researched to prevent floods and droughts are to plant trees, to build large-scale forests nationwide. The temporary solutions to these disasters involve using machinery to pump water and construct high levees and clear and deepen river channels. If these two methods, addressing the symptoms and the root causes, can be fully implemented, then disasters like floods and droughts can be avoided, and thus, the production of food will not suffer losses.[37]

Sun's emphasis on forestry was grounded in his concern both for individuals' survival and for the path of the nation. By pointing out the role the failure of food production played in affecting the Russian Revolution, he expressed his anxiety about the natural calamities that could result in social unrest. For Sun, to safeguard food production and ensure social stability, forestry should no longer be considered subordinate to agriculture since the former protects the latter.

Sun's exposition represents a trend toward highlighting forestry in the agriculture-based Chinese society. In *The Rural Modern*, Kate Merkel-Hess shows how intellectuals in the Republican era popularized modern science among the rural masses.[38] When Ling Daoyang and Sun Yat-sen promoted afforestation, they were actually developing a new vision for solving the problems of drought and flood in rural China, an effort that was equally as, if not more, important than promoting literacy and self-governance organizations. *Conquering Nature* (*Zhengfu ziran*), published by the Chinese Civilian Education Promotion Association (Zhonghua pingmin jiaoyu cujin hui) in 1932, expresses this necessity of transforming the natural environment through tree planting (see figure 1.3). Launched by the famous educators Tao Xingzhi and Yan Yangchu, the association has aimed at transmitting scientific knowledge to the rural inhabitants and improving their daily lives using easy-to-implement practices. *Conquering Nature* argues that cactus and other drought-resistant plants can, over time, change hardscrabble ground into flourishing forestland: "Improving barren stony land in nature is an example of human beings conquering nature."[39] Adopting the rhetoric of Ling Daoyang and other foresters, the book claims that trees can increase rainfall and adjust moisture. *Arbor Day* (*Zhishu jie*), another book published by the Chinese Civilian Education Promotion Association, encourages farmers to plant fruit trees to improve their livelihoods.[40] This imperative of feeding the rural peasants echoes Sun's efforts to build a strong, prosperous, and self-reliant nation.

For Sun Yat-sen, among solutions that also include transportation, machines, fertilizers, and pest control, promoting forestry in rural China is the most important way of dealing with food shortages. Tree planting can address the root causes of flooding and droughts once and for all, but more importantly, it can ensure the stability of the society. Nationalism thus coalesces with ecological exigency.

Sun's commitment to forestry included not only planting trees but also sustainably harvesting them. In "Plan for Building the Nation" ("Jianguo fanglüe"), he argued that cutting down primeval forests in the northwestern part of the country is justifiable on the grounds that the nation should have its own paper industry.[41] He lamented that China, which claims to have invented paper in ancient times, must now import it. Sun's goal in promoting self-sufficiency with respect to paper production is associated with

FIGURE 1.3 **Cover of *Conquering Nature* (*Zhengfu ziran*), sponsored and published by the Chinese Civilian Education Promotion Association in 1932. Source: Zhou Hui'an, *Zhengfu ziran*.**

his desire to educate the common people through more easily accessible printed products, including newspapers, encyclopedias, and translations of books from Western countries. A tree, for Sun, is not only a means to achieve economic independence but also a social-engineering medium to make the Chinese better educated and more closely connected to each other in their hearts and soul.

Sun's desire to encompass both planting and harvesting trees in his forestry scheme is not inherently contradictory because his goals were not limited to environmental conservation but also included developing the nation-state. He also encouraged the exploitation of other natural resources, including bronze, gold, and oil.[42] For example, he urged the mining of bronze in Yunnan and Sichuan so that the nation would not have to import it. Discussing the role of coal in reframing China's status in the world, Shellen Wu argues that mining has been inextricably linked to China's sovereignty and state power since the late nineteenth century.[43] Likewise, Sun endeavored to set forestry as a pivotal component of nation building, suggesting that forestry is essential to enhancing China's economic and political status on the global stage.

Sun's agenda to both plant and cut down trees was driven by China's dire financial straits and the resulting social problems. In his treatise on railroad construction and the people's welfare, he argued that forestry, together with railroads, tap water, electricity, gas, trolley cars, and canals, should be managed by the government.[44] The government should tax most natural resources, including timber, to support education, the elderly, and other social-welfare purposes, he claimed. Though Sun did not specify how the nation could profit from forestry, he clearly intended for it to be used to generate revenue sources to help the poor.

Also in the vein of improving the welfare of the common people, Sun called for supporting workers' interests, asserting that forests should not fall into the hands of rich businessmen. He argued that in Europe, tycoons have monopolized natural resources, provoking uprisings by the socialists, who destroyed the railroads and other means of production. Therefore, it is the government's responsibility to make natural resources accessible to the public.[45] Sun wanted the common people, rather than the businessmen, to dominate the natural resources.

In 1923, Sun's vision of forestry was encoded into the Nationalist Party's official doctrines; this was a huge leap since it meant that tree planting was becoming institutionalized. That year, a declaration of the Nationalist Party stipulated that the forests were the property of the people and would be managed by the government.[46] Sun's stance coincided with the actions of US president Theodore Roosevelt, who, at the beginning of the twentieth century, set aside public land for national parks to be enjoyed by the public.[47] Having been to the United States during that period, Sun was possibly influenced by Roosevelt's pioneering efforts in connecting public interests and environmental conservation. In 1924, in a speech to the first Nationalist congressional National Assembly, Sun again called for the government to play an active role in forest management as a way to shore up education, health, and other social-welfare benefits.

Arbor Day's Symbolic Twist: The Cracks in China's Modern Environmentalism

While Sun Yat-sen strove to show that a modern nation cannot survive and flourish without a well-developed forestry plan, after his death on March 12, 1925, his followers paradoxically marginalized his tree-planting endeavors through changes in the celebration of Arbor Day.[48] As Arbor Day gained traction, the act of tree planting took on symbolic and political significance. This shift paved the way for the emergence of two distinct forms of environmentalism—actual tree planting and symbolic gestures—both of which became increasingly contested. As Arbor Day became more ritualized, it gradually departed from its original focus on tree planting. The ritualization of Arbor Day signals the decline of China's environmentalism under the Nationalist government.

People began to reflect upon the real outcomes of Arbor Day in the mid-1920s. An article entitled "Suggestions to Rectify Arbor Day" ("Duiyu zhishujie yingxing zhengdun zhi yijian") in a newspaper in Jiangsu Province argues that the local residents had underestimated the size of the barren land and lacked a survey of the geographical features and suggests that it was not realistic to plant trees on such a large scale.[49] The timing of the holiday also became an issue: another article points out that Arbor Day is usually sunny, which is detrimental to seedlings' growth; therefore, a day in the rainy season would be optimal.[50] A journal for farmers points out that the United States had not set a date for Arbor Day nationwide but had left to each state to choose its own date according to local circumstances.[51] The article suggests that local farmers should adjust the date of Arbor Day to ensure the survival of newly planted trees.

Concerns regarding the effectiveness of tree planting gained traction in the late 1920s. In the 1926 novella *Arbor Day* (*Zhishujie*) by Zhang Ziping, the protagonist, known as V., is a poor schoolteacher struggling to make ends meet. During the Arbor Day holiday, which coincides with the Qingming festival, he reflects on his frustration with teaching and guilt about not being able to provide a decent life for his family. At the end of the novella, he goes into a store where both rice and coffins are sold and is overtaken by a spell of madness: he wonders which he should buy.[52] The work is largely symbolic; on the surface, it explores rising inflation and the hardships of the urban poor. The author suggests that victory—the full form of the protagonist's name—can hardly be achieved when the citizens are not mobilized to plant trees. During Arbor Day, V. is struggling to choose between rice, which embodies agricultural production and survival, and the wooden coffin, which symbolizes death. Using "Arbor Day" as the title, the novella implies a possible solution to V.'s conundrum: tree planting. The agricultural products safeguarded by the ecological roles of trees, on the one hand, and the optimistic associations of trees, on the other hand, will rejuvenate the impoverished poor and serve as the antidote to V.'s physical, financial, and mental suffering. The novella *Arbor Day* shows that Arbor Day should not be merely a date on the calendar but a day for real action. In 1928, a Shanghai-based newspaper argued that the tree-planting campaign during Arbor Day had become more of a symbolic exercise than a real action. Citizens attended the movement only passively and did not acquire the necessary forestry knowledge and mindset to conserve trees in their everyday practice.[53]

In 1928, the spring equinox on March 21 replaced the Qingming festival as the date of Arbor Day as it is the optimal time to plant trees in Zhejiang.[54] In Shanghai, Arbor Day was set on March 12 to memorialize Sun Yat-sen.[55] To inherit the legacy of Sun Yat-sen and to leave behind the old tradition of tree planting from the era of Yuan Shikai and the later Beiyang regime, establishing a new Arbor Day date seems to have been inevitable.[56] In 1929, Arbor Day was embraced as a national celebration throughout most of the territories governed by the Nationalist regime, marking the anniversary of Sun Yat-sen's passing by observing this special day on March 12.[57] As the Nationalist government expanded its control over a broad expanse of China, the nationwide observance of Arbor Day fostered cohesion and established a sense of steadiness. While setting up a new date, both environmental constraints and traditional culture were taken into account. The period around Sun Yat-sen's death date was chosen because it was warm enough for tree planting in Nanjing. Arbor Day later adopted the pathos of the Qingming festival, which traditionally memorializes the deceased, to specifically honor Sun Yat-sen.

Arbor Day fuses the past and the future into a single moment of reflection and promise. Yet it is this very integration that has caused a rift in its celebration. As a day that aspires both to secure the prosperity of future generations and to honor the legacy of a pivotal historical figure, it necessitates the inheritance of cultural capital from the past. But the significance of this legacy is intricately tied to the ever-changing political landscape of the present, dictating its worth and relevance. After the Nationalists unified the vast territory of China in 1927, the government increasingly needed to deploy a political symbol to prove its legitimacy and exercise its power nationwide. Tree planting had served the purpose of celebrating Sun Yat-sen's political and cultural assets by his heirs, who were eager to compete for power in the shifting tides of the political landscape. Somewhat paradoxically, the memorialization of Sun's past endeavors hindered as well as advanced the prospects of Arbor Day.

Sun's promotion of material prosperity by tree planting first began to materialize in China's most affluent city, Shanghai, where the benefits of trees were perceived mainly from an economic perspective. One Shanghai newspaper article, for example, proclaims that tree planting is the best form of savings. The article asks citizens to "deposit trees in the bank of conservation."[58] It argues that planting trees is like contributing to a bank account whose earnings far exceed the principal investment. And compared with actual bank dividends, the earnings from the trees are more reliable because the savings in the bank may depreciate or evaporate in bankruptcy, while the value created by trees will grow continuously. In twenty years, they will be worth a considerable amount of money. As China's richest citizens, the article goes on to say, Shanghai residents should contribute either their money or their labor to tree planting because it improves people's livelihoods, as Sun had advocated.

Despite having been promoted in China's largest city, the final outcome of tree-planting efforts during Arbor Day was not encouraging. The fact that the public needed continual reminders from the local media suggests that the effort to motivate the populace was far from easy. A 1933 editorial complains that Arbor Day is only a formality and a waste of money.[59] A newspaper report claims that Arbor Day is merely symbolic because the trees were not taken care of after being planted and photographed on Arbor Day and points out that no more than 20 percent of the trees survived.[60]

Planting trees became increasingly ritualized, adding more political significance to the practice. Beginning in 1929, the tree-planting ceremony was carefully designed to accommodate political needs. It involved ten steps: first, the playing of music; second, standing in respect; third, saluting Sun Yat-sen's photo and the national flag; fourth, reading Sun's will; fifth, standing in silent tribute for three minutes; sixth, the chair of the conference reading the goals of the event; seventh, a speech; eighth, singing songs and planting trees; ninth, repeating slogans; and tenth, taking photographs.[61] The ceremony endowed tree planting with new political meanings, including loyalty to the nation-state and keeping Sun's memory alive. As historian Henrietta Harrison explains, "Rituals and ceremonies are a common way of creating or imposing this kind of [ethnic and national] classification which provides a vision of the social world and of establishing a consensus about its meaning."[62] The significance of Arbor Day thus lays not in its ecological concerns but in the ceremony itself as it was used to cultivate ties between citizens and the state.

To be sure, from the outset, Arbor Day had conveyed meanings beyond mere tree planting. First celebrated in the United States, Arbor Day was imbued there with economic, environmental, and cultural connotations. A number of environmentalists collaborated to commemorate the first US Arbor Day. American environmentalist George Perkins Marsh had been impressed by European efforts in planting trees and setting tree planting as a profession in colleges. In his 1864 masterpiece, *Man and Nature,* he reacted to the rapid depletion of forests in the United States and tried to resolve it through afforestation.[63] He argued that trees can "diminish the frequency and violence of river inundations, prevent the formation of new torrents and check the violence of those already existing, mitigate the extremes of atmospheric temperature, humidity, and precipitation . . . furnish a self-renewing and inexhaustible supply of material indispensable to so many purposes of domestic comfort."[64] In 1872, J. Sterling Morton (1832–1903), who served as US secretary of agriculture from 1893 to 1897, proposed Arbor Day in Nebraska to institutionalize tree planting. Two years later, Nebraska governor Robert W. Furnas issued the first proclamation of Arbor Day. In 1885, the day became a legal holiday when the legislature set aside April 22, Morton's birthday, as Arbor Day. Other states soon followed suit, and Arbor Day spread across the country. In the United States, Arbor Day was meant to educate schoolchildren about trees, nature, and conservation as well as to promote good citizenship.[65]

In examining the ways sylvan life might affect the human mind, Morton interpreted the tree as the "Tree of Knowledge," the "Tree of Liberty," and the "Tree of Everlasting Life."[66] In so doing, he integrated conservationist ideas and egalitarian and democratic values and disseminated them to Americans. In China, on the other hand, Arbor Day has manifested and intensified political hierarchy. Henrietta Harrison contends that "the rituals of the Chinese republic created communities, but they also created hierarchies of power between those communities."[67] A 1930 report on planting trees along the railway is exemplary of this power dynamic.[68] In the first image in the report, a senior official demonstrates to his subordinates how to plant trees (see figure 1.4); they show their loyalty by learning from him. After the tree has been planted, the official appears proud as he stands in front of the camera (see figure 1.5). Although the ceremony is intended to celebrate the fifth anniversary of Sun's death, it is more a construction of a leader's authority than a campaign to green the nation—no mention is made of environmental concerns.

The rivalry between the Nationalists and the Communists as well as the

FIGURES 1.4 **A senior official plants a tree on Arbor Day as his subordinates look on. Source: *Nanxuntieluyuebao*, March, 1930.**

FIGURES 1.5 The official proudly stands in front of the tree he has planted during Arbor Day. Source: *Nanxuntieluyuebao*, March, 1930.

imminent threat from Japan were also drawn into the symbolism of Arbor Day. In 1931, Chiang Kai-shek participated in the deceased soldiers' memorial ceremony on the sixth anniversary of Sun's death.[69] Chiang was particularly anxious to contain the Communists. At this ceremony, he barely mentioned the environmental or economic benefits of tree planting.[70] Instead, planting trees was framed as a means of reducing the trauma of war and evoking patriotism.

Even though environmental concerns were sometimes mentioned in the media, tree planting was often politicized. A 1931 journal article entitled "Documenting Tree Planting" ("Zhishu ji") promoted the ecological benefits of planting trees, including cleaning the air and conserving water. Nevertheless, the author claimed that the goal of Arbor Day is "not only to commemorate Sun but his invincible spirit for revolution during these forty years" and "not only to plant trees but commemorate [Sun's] various plans to seek the security of the world and prosperity of the people."[71] The newspaper narrowed the purportedly cosmopolitan goal of global security to local clashes between the Nationalists and the Communists. The author said, "I am only worried about the bandit Communists, who disrupt things everywhere, leaving the premier's revolution unfinished."[72] Tree planting was first universalized and sublimated as a desire for peace and then relegated to a narrow political task. Communists were criminalized for not following Sun's instructions. The author employed Sun's rhetoric but reframed it for his own purposes. On the one hand, the article highlighted Sun's concern for national survival. On the other, this celebration of Sun was detached from Sun's plans for ecological conservation and economic development.

While Arbor Day was becoming politicized within the Nationalist Party, the clash between two self-proclaimed followers of Sun, Chiang Kai-shek and Wang Jingwei, shows how Arbor Day could be manipulated to support a government's political legitimacy. Larissa Pitts argues that Arbor Day established a sense of unity by encouraging individuals to imagine themselves as part of a larger national community.[73] In studying the Beiyang government from 1915 to 1927, Pitts suggests that celebrating tree planting on a specific day served as a vehicle to cultivate political participation among the Chinese population as a whole. I suggest that if we extend our timescale to 1949, it is more illuminating to see that Arbor Day symbolized and intensified political contestation within China. The clash between Chiang and Wang reveals that modern environmentalism was beginning to crack and even fail when Arbor Day became a contested rhetorical arena instead of an actual tree-planting practice.

In 1934, Chiang made tree planting an essential component of his New Life movement to cultivate good citizenship and promote civic responsibility. The movement called for taking a fresh look at everyday hygiene, order, and morality and aspired to reform the Chinese character and mind by inculcating new habits and mobilizing individual citizens amid national crises. In the same year, Chiang declared that he would launch a new movement from Jiangxi, the ideal location for eliminating the Communists and providing the foundation for a national revival. He urged citizens to change their daily behaviors, including ceasing spitting and smoking and starting to wash their faces with cold water. It is through these small actions, he contended, that the Chinese would become civilized and stronger.[74] The movement also encouraged planting trees in spring, alongside cleaning in summer, saving money in fall, and subsidizing the weak in winter.[75]

Influenced by the military discipline of Germany, the New Life movement borrowed the German belief that interaction with the natural environment helps shape a strong national character. The German environmentalist Konrad Guenther claims that "the source of Germanic national character was inexhaustible . . . because it lay in the darkness of the forests, where it was beyond the reach of enemy hands."[76] Inspired by this thought, Germany initiated a massive tree-planting project. For Chiang and his subordinates, tree planting could cultivate a rigorous and optimistic mindset similar to that of Germany.[77] Maggie Clinton explains that the New Life movement was modern, militarily ordered, and rigidly hierarchical, with a relentless emphasis on national unity and cohesion.[78] Wearing his military uniform, Chiang himself participated in tree planting in Nanchang and called on his citizens to follow his lead. He hoped that a more disciplined citizenry would result.[79] Tree planting was thus a way to exercise his political leadership and reinforce social hierarchies.

The involvement of children in Arbor Day during the New Life movement not only introduced knowledge of nature originating from foresters but also instilled collective power and discipline in the younger generation. In 1936, a textbook for primary school students published in Shanghai called on children to plant trees in order to prevent floods like the one that had brought about massive damage in 1930. The book, which documented how planting trees could bring about direct as well as indirect benefits, echoed the arguments of foresters in earlier decades.[80] In a lesson on Arbor Day in *Reviving the Chinese-Language Textbook* (*Fuxing guoyu keben*), a child plants trees with his mother in their backyard.[81] After they finish, he sings tree-planting songs. The lyrics include "A tiny tree this morning stands / Years hence, a hundred feet it spans" (Jinzhao xiao xiao yi zhu shu / Ta nian gao gao bai chi chang). The lyrics express and create a feeling of sympathy.[82] The textbook encouraged children to identify with the trees they planted, to value the long-term investment that accompanies the life of a healthy tree, and to take pride in the stewardship that the planting embodies. It then introduced what children can learn after the trees have grown up. After the lesson on Arbor Day, the next one depicted squirrels in the forest. The squirrels are sensitive to human beings and clever in the way they store food. Planting trees thus became an entry point for children to understand the natural world.[83] An article in the magazine *Children's World* (*Ertong shijie*) claimed that children can learn the basic principles of nature by planting trees.[84] Every child participating planted ten trees, and together, they successfully planted a total of three thousand trees. If only one person had undertaken this task, it would have taken him or her over ten days to achieve the same result. The article further suggested that Arbor Day provides a chance for children to socialize and promotes collective power and discipline.

The contestation of Chinese environmentalism became more dramatic when Arbor Day served as a point of contention between Chiang Kai-shek and Wang Jingwei. Like Chiang, Wang memorialized Sun Yat-sen by planting trees during the Second World War. Supported by the Japanese, Wang headed a new government in Nanjing that opposed Chiang's government in Chongqing. Like Chiang's New Life movement, Wang's New Citizen movement included planting trees, aiming to reshape the public's everyday life and revive the morality of degenerate citizens.[85] Wang sent his representatives to participate in tree planting at the Sun Yat-sen Mausoleum in Nanjing in 1942, the seventeenth anniversary of Sun's death (see figure 1.6). On the same date, Chiang's government also celebrated Arbor Day in Chongqing. The 1942 Arbor Day celebration coincided with and was integrated into Chiang's wartime spiritual mobilization movement, aiming to cultivate a healthy and combative mentality.[86]

Why did both Chiang's and Wang's regimes endorse tree planting in an era of upheaval? As Sun's followers, Wang and Chiang waged a symbolic war as they participated in tree planting.[87] In doing so, they aspired to portray themselves as disciples and faithful guardians of Sun's memory. As a result, planting trees was not only politicized but also potentially weaponized to delegitimize opponents.

The clash between the governments of Chiang and Wang for political legitimacy through tree planting illustrates the fluidity and adaptability of environmental initiatives. As citizens plant trees with their leaders, they implicitly acknowledge the leaders' legitimacy. For these two leaders, political legitimacy was achieved by daily practice rather than by issuing political

FIGURE 1.6 **Representatives of Wang Jingwei's government participated in planting trees at the Sun Yat-sen Mausoleum in Nanjing in 1942. A tablet says, "Memorial forest [*jinian lin*] for the seventeenth passing anniversary of the Father of the Nation." Source: Beijing University Library, public domain.**

manifestoes. Environmental initiatives, on the one hand, promote conservation and social welfare. On the other hand, in enforcing environmental initiatives, leaders invisibly and strategically add new political intentions and eventually change the original goals of environmental conservation. As will be seen in chapter 3, this camouflaging of political intentions by environmental activities became more confrontational during the Mao era than that of Chiang and Wang.

The participation of the Japanese in Arbor Day during the Second Sino-Japanese War added another layer of symbolism. In 1943, *Shanghai News* (*Shenbao*) reported that officials and young students from both Japan and Japanese-controlled China celebrated Arbor Day in Nanjing and Wuhan. They memorialized Sun Yat-sen by planting trees as well as visiting his bronze statue.[88] The article suggested that Japanese youth in Nanjing admired Sun Yat-sen's initial efforts in Japan advocating for Pan-Asianism.

Keen on extending their sway in China, the Japanese aimed to strengthen Sino-Japanese ties by highlighting Sun's connections to Japan. Arbor Day's memorializing of Sun offered the Japanese an opportunity to infiltrate into Chinese cultural exercises in a less visible manner. While the ecological and economic benefits of trees fell into oblivion, celebrating Arbor Day further served as a means to demonstrate strong solidarity between the Chinese and Japanese as the Japanese also paid tribute to Sun. Arbor Day evolved into a symbol of Pan-Asianism, calling for cooperation and unity between the Chinese and the rest of Asia.

Despite the observance of Arbor Day during the Second World War, the celebration of tree planting did not translate into tangible efforts to plant trees in China. In his *The Omnipotence of Forests* (*Senlin wannenglun*), forester Hao Jingsheng claimed that forestry has never been seriously taken into account. He further argued that after the War of Resistance against the Japanese, China should catch up with the West immediately.[89] In the opening chapter, he imagined a debate between two protagonists, Mr. Steel and Ms. Forest. Mr. Steel is a domineering figure, serving as the pillar of the country, but Ms. Forest is his equal in promoting national development. Gendering the forest as female sheds light upon a little-noticed but indispensable role of the forest for nation building. Hao noted that during the Second World War, Germany used wood to produce combustible gas to power its tanks. In addition, wood fiber was essential in making clothes, shoes, and food. From soldiers on the battlefield to civilians at home, Hao suggested, everyone was indebted to Ms. Forest. Like the foresters of the 1920s, Hao, in 1944, was still promoting forestry as a distinctive discipline, one that permeated and improved every aspect of individuals' daily lives. Almost three decades after the first observance of Arbor Day, afforestation was still a rather novel project that had yet to reach fruition on a national scale.

During the Chinese Civil War between the Nationalists and the Communists (1945–49), tree planting became increasingly marginalized. In 1948, Hao Jingsheng noted that China's forestry services accounted for only 0.04 percent of the national budget. Because of constant war and inflation, this small investment could hardly realize the country's goals of afforestation.[90] That same year, forestry and agriculture minister Zuo Shunsheng stated that it was hard for the common people to plant trees, let alone protect them, because they lacked basic knowledge of forestry.[91] China still had not established any scientific research on forestry, which caused it to fall behind the West. Zuo argued that forestry could promote the livelihood of the people, in contrast to the actions of the Communists, who were only causing chaos. Zuo's argument shows how tree planting had become a symbolic gesture designed to denigrate the Communists. Ordinary citizens also noticed the symbolic nature of tree planting. A writer in Shanghai complained that the land was still not covered by trees and that the seedlings died without special care, though tree planting had been advocated for twenty years.[92] In Guangzhou, another writer lamented that seedlings were not properly taken care of and that constant war had devastated the forest.[93]

On the eve of the Nationalists' defeat in 1949, He Yingqin, the minister of defense, celebrated his sixtieth birthday during Arbor Day. His smiling face was featured on the cover of *News World* (*Xinwen tiandi*), an influential magazine widely read by politicians, intellectuals, and celebrities in the Republican era, while images of trees were entirely absent.[94] Sun Yat-sen's earnest call for planting trees had been transformed into a merely symbolic gesture.

The Seeds of Red Environmentalism: Communists' Tree Planting without Arbor Day

The contestation of modern Chinese environmentalisms becomes more acute if one considers the environmental initiatives of the Communists. While the Nationalists' efforts to plant trees became increasingly symbolic, the Communists emphasized the concrete results of tree planting and endeavored to transmit recognition of the benefits of forestry to the masses. While the Communists did not celebrate the legacy of Sun Yat-sen during Arbor Day celebrations, they were indebted to the Nationalists for their forestry knowledge. Exploring the Communists' pre-1949 environmental ideas, which have received relatively little attention, provides an intriguing glimpse into the evolution of environmentalism among them. Additionally, it reveals how these concepts were disseminated and adapted to different social and political circumstances.

The educational background of Jiang Xin, a first-generation forester in Communist China, reveals that the Communists' forestry knowledge came from schools run by the Beiyang and, later, the Nationalist governments. Educated as a forestry major at the Second Vocational School of Shanxi Province in 1924 and at Shanxi Agricultural Specialized School in 1928, Jiang Xin was indebted to Ling Daoyang, though they may not have met each other.[95] Originally set up as Shanxi Agriculture and Forestry School in 1902 by the late-Qing governor of Shanxi, Ling Chunxuan, Shanxi Agricultural Specialized School was one of the first to specialize in forestry in the Republican era.[96] Though China was suffering from warlord conflicts, multiple factions seemed to reach a consensus on highlighting forestry, which indicates that the benefits of trees and forests had gradually come to be appreciated by a divided society.

Tracing the educational background of Jiang Xin reveals not only the transnational scope of forestry knowledge in Republican China but also the genealogy of the Communists' first-generation foresters. Ling Daoyang and other foresters learned their science from the United States and popularized it during the Beiyang and, later, Nationalist governments. Educated in the forestry school that used these Western-educated foresters' textbooks, the first-generation Communist government's foresters were more the offspring of the West than of the Soviet Union.[97] Still, the Communists' transmission of forestry targeted the peasants—in sharp contrast to that of the Nationalists, whose symbolic celebration of Arbor Day marginalized the ecological and economic benefits of tree planting.

The earliest form of the Communists' tree-planting exercise can be traced back at least to the 1930s. As early as 1932, the Communist Party newspaper *Red China* (*Hongse Zhonghua*) called local cadres to launch a tree-planting movement in Jiangxi Province, which was the political center of the Chinese Soviet Republic from 1931 to 1937. The provisional central government of the Chinese Soviet Republic passed an act calling on local officials to teach farmers the benefits of forestry and trees.[98] In 1934, one newspaper article argued that a massive tree-planting campaign should be launched immediately: first, the central land department should ask local leaders to form a tree-planting committee; next, the people should be fully mobilized and instructed in how they can benefit from tree planting. The article elaborated on the technical details of tree planting, including sowing, cutting, and transplanting seedlings, hoping to teach the local farmers how to successfully plant a tree.[99] To protect mountain trees, the Communist regime enacted a law: those caught cutting down trees or setting fires to clear land would face fines, forced labor, and imprisonment.[100] In April 1934, a landlord was executed for setting a massive fire that destroyed a large area of forest.[101]

After the conclusion of the Long March, Mao arrived in October 1935 in Yan'an, which was plagued by soil erosion and low rainfall, and the survival of tree-planting efforts became a more urgent issue. In 1942, Jiang Xin published an article, "Tree Planting and Forest Protection" ("Zhishu he hulin"), in the Communist newspaper *Liberation Daily* (*Jiefang ribao*). He argued that Arbor Day had become "Dead Tree Day" because of the failure to take care of the seedlings after planting. He suggested that specific officials should be assigned to ensure the survival of the trees. He asserted that "if you don't have confidence [in their survival], you should not plant trees."[102]

Jiang integrated daily economic needs and long-term ecological concerns to illuminate the benefits of tree planting. He pointed out that the masses' daily necessities depended on trees: doors, window frames, desks, and chairs are all made from wood. In addition, the ecological role played by trees is vital to the people's survival. Jiang noted that in the very northern part of Shaanxi, famine was worse than in Yan'an because Yan'an had more forests. Meanwhile, he maintained, the desert was encroaching, threatening to engulf the entire region. If the government wanted to thwart the oncoming desertification, it had to recognize the urgency of planting trees.

For the Communists, tree planting safeguarded the agricultural economy in Yan'an. At the beginning of the 1940s, famine, drought, and frost plagued the Yan'an region. A small drought took place every five years; a larger one, every ten years.[103] Tree planting not only was an essential means to boost agricultural production but also provided daily necessities. In 1941, the value of forest products—including timber, food, and seeds—reached almost seven and a half million yuan every year.[104]

While Nationalists Chiang Kai-shek and Wang Jingwei were competing for political legitimacy in deploying the symbolism of Arbor Day in the early 1940s, the Communists took pains to instruct the masses on how to plant trees properly. A 1943 article in *Liberation Daily* introduced several key aspects of tree planting: when to plant, how to nurture the seedlings, techniques for cultivating pear and apple trees, and pest control.[105] For example, a seedling should not be transplanted until it is one year old. When the seedling is transferred from one pot to another, its topsoil should go with it. These details show that the Communists were deeply committed to the success of tree planting and tried to persuade the peasants to plant trees scientifically and effectively; doing so would result in mature forests and orchards rather than planting seedlings in a largely symbolic gesture only to let them wither and die, as the Nationalists had done.

As the Second World War escalated, the Communists prioritized trees' benefits for military purposes (this subject will be explored in more detail in chapter 4). In a 1944 article, Jiang Xin argued that trees were associated with the war against the Japanese at the time.[106] He claimed that modern war is a three-dimensional exercise and forests provide excellent camouflage and thus protect against Japanese airplanes. If an area is covered by forest, it would be hard for the Japanese to find the Chinese bases and arsenals. Also, if the forest canopy is large enough, it would be difficult for the enemy to identify military goods and personnel being transported by rail. If the Japanese use chemical weapons, tree trunks could inhibit their spread; also, soldiers could save their lives by climbing the trees because the fumes disperse on the ground. Furthermore, wood could be used to produce activated carbon to absorb the chemicals. In addition to providing defensive resources, Jiang suggested, trees could be used to produce offensive military equipment, such as airplanes, rifle butts, and railroad ties. And processed wood is integral to the production of dynamite. In addition to military uses, forestry is a must for civilian welfare. For example, trees could be made into artificial silk to replace clothes that have undergone chemical reactions.

Like Ling Daoyang, Jiang Xin saw the ecological benefits of trees as indirect. Jiang elaborated on the five ecological roles of trees: fighting floods, preventing famine, stopping the desert from spreading, preserving topsoil, and moderating the climate.[107] Claiming that rainfall and sand are the two main reasons for floods, Jiang argued that tree planting could solve these problems (a similar contention by the Mao-era forester Liang Xi will be examined in the next chapter). When a river cannot accommodate a heavy rain or if sand makes up the riverbed, floods are likely. Trees offer a cushion against rain, which cannot directly penetrate the ground because the trunks and branches slow it as it falls. When the rain finally reaches the ground, roots can absorb it. When there are more trees around the banks of a river, the river carries less silt because the trees help prevent soil runoff. Specifically, in Nanniwan, where the Communist government boasted of its success in fighting against the harsh natural environment, Jiang warned that the loss of trees may ultimately devastate the region.

While Jiang's exposition, in many ways, is reminiscent of Ling Daoyang's *Main Ideas of Forestry*, the two works differ in their time frames. While Ling called for long-term investment in trees for their many benefits, Jiang's concerns were more immediate, as when he elaborated on how trees can contribute to battle strategies. While Ling engaged primarily with intellectuals, scientists, and policymakers, Jiang was addressing cadres and the masses who were tackling ecological problems in their daily lives at the time. Thus, Ling provides a wide-angle national picture of the benefits of trees, while Jiang's approach focuses more on how trees could contribute to improving local circumstances, especially during wartime.

A *Liberation Daily* article regarding tree planting in the Communist capital, Yan'an, makes it clear that popularizing forestry knowledge was a top priority in wartime. The Communists' endeavor to promote tree planting continued during the Civil War that followed the Second World War. While the article suggests that the Communists might not succeed in ensuring the survival of seedlings, it also reveals that they had been concerned about the actual survival of the trees rather than anything merely symbolic. To ensure success at the grassroots level, Jiang argued, local government should provide financial support to farmers, for whom trees can bring about material benefits: "Planting one more tree means feeding one more sheep."[108] In addition, drama, pamphlets, folk dance, and official newspapers could mobilize peasants to take up tree planting. Jiang claimed, for instance, that the peasants did not know that balsam poplars cannot be grafted. They lacked the habits of proper tree planting owing to a deficiency in forestry knowledge. Some of the cadres apparently did not even know that trees grow from seeds. Thus, the article said, it was urgent that forestry knowledge be popularized among the peasants and cadres.

Instead of seeing tree planting as just a symbolic gesture, then, the Communists had the patience to ensure the survival of seedlings.[109] While the Communists did not celebrate Arbor Day—which was then widely observed by the Nationalists—they nonetheless had their own massive tree-planting initiatives that led to the same outcomes. When the Communists called on people to plant trees during the Qingming festival, they were accommodating locals' needs. Different locations could choose their own policies according to individual circumstances. They could, for instance, select tree species based on the particularities of the region. To encourage the masses to participate in tree planting, the local government selected model workers to share their successes. One such model worker claimed that "if all plant trees, all would have moneymaking trees [*yaoqianshu*]".[110]

In addition to inculcating knowledge of trees among adults, the Communists, like their Beiyang and Nationalist predecessors and contemporaries, instructed children to plant trees. A 1948 textbook said that trees can provide food, produce materials to build houses, and prevent windstorms and floods. At the end of the lesson, because of the benefits of trees, a child says that he must plant three trees and keep them alive this year.[111] Compared with the textbook promoted by the Beiyang and Nationalist governments, which introduced the natural environment and then called on children to socialize during Arbor Day, the Communist book suggested that keeping the trees alive was the top priority for the children.

While, as noted above, the Communists did not celebrate Arbor Day, their tree-planting efforts were heirs of Ling Daoyang and Sun Yat-sen in the Republican era. Ling's knowledge of forestry, in turn, came from the United States, which derived it from Germany and France. The Communists integrated this global knowledge of forestry into their agricultural practices in a desperate time, advocating real tree planting rather than transforming the practice into a symbolic gesture. Somewhat ironically, then, it was the Communists who did not celebrate Sun Yat-sen or Arbor Day who carried on Sun's legacy by mobilizing people to plant trees as part of their daily lives.

Beyond 1949: Arbor Day Lost, Arbor Day Regained

Tree planting took a dramatic turn after 1949 under both the Nationalist and the Communist governments. In 1949, the Nationalists retreated to Taiwan, the new home of the Republic of China, where they continued to practice tree planting in memory of Sun.[112] In a 2016 article, a century after the first Chinese Arbor Day, Guan Bingzong, a professor of forestry management in Taiwan, asserted that the island should not bring in timber from overseas; he was alerting his readers to the threat of excessive reliance on timber imports when other nations had been unfriendly to Taiwan's international trade.[113]

For the Communists, Sun Yat-sen's ideas of planting trees to save the nation in crisis survived, though his name was mostly left off. Across the Taiwan Strait, after coming into power in the mainland in 1949, the Communists glossed over the endeavors of Sun and foresters in the Republican era. In the Mao era, the Communist mouthpiece newspaper the *People's Daily* (*Renmin ribao*) portrayed life in the Republic of China as a period of deforestation and exploitation of cheap labor under Chiang Kai-shek's brutal leadership. It dismissed Arbor Day as a symbolic gesture for the bureaucrats to memorialize Sun Yat-sen, thereby turning it into a "forest destruction day" (*senlin pohuai jie*).[114] In practice, Sun's rhetoric of tree planting was lost to the Communist government's cultural propaganda as well as its social and political agenda (see figure 1.7). Sun would probably not have anticipated that the party he had tried to cooperate with would eventually marginalize his endeavor.

In 1979, Arbor Day officially resumed in mainland China on the anniversary of Sun's death, but his pioneering endeavor and the reason for the choice of date were hardly mentioned; the symbolic meanings of Arbor Day became selective. On April 3, 2020, the embattled party leader Xi Jinping planted trees in Beijing when China was widely accused of negligence during the outbreak of the COVID-19 pandemic.[115] By planting trees, Xi suggested that he was still in charge and that China was on its way to recovery—a practice that is redolent of Chiang and Wang, who used tree planting as a symbolic gesture to demonstrate their leadership and confidence in difficult times.

FIGURE 1.7 **Arbor Day–themed stamps issued on March 12, 2019. That the date is Sun Yat-sen's death anniversary is not mentioned. Photo by the author.**

Conclusion

This chapter has embarked on an exploration of the intricate and contested domains of environmentalisms, utilizing Arbor Day and its cultural relevance as a focal point. The interplay between tangible tree-planting efforts and symbolic gestures in Republican China serves as an embodiment of the larger dynamic between the yearning for a deeper connection with the natural world and the compulsion to establish social networks—an interplay that propels the fluctuating trajectory of contemporary environmentalisms in China.

The founding of Arbor Day was grounded in foresters' desire to naturalize an environment ravaged by social unrest and war. Foresters convinced policymakers of the critical role of forestry in nation building. They viewed the economic benefits of trees as the direct ones, while classifying the ecological benefits as indirect. In the case of the former, they strove to promote the role of forestry in fostering national survival. In regard to the latter, they engineered a new mode of seeing nature. Modern forestry has offered a scientific interpretation of natural changes that is applicable everywhere. This scientific rationalism makes nature more comprehensible, calculable, predictable—and profitable.

Arbor Day became a means of cultivating modern citizens. During the New Life movement, planting trees functioned as an exercise to promote healthy living habits. Engaging in the practice also mobilized citizens, including children, to become more socialized and disciplined. As the war between the Nationalist government and the Communist revolutionaries escalated, for the former, Arbor Day increasingly became either a symbolic exercise to affirm the political legitimacy of Chiang Kai-shek and Wang Jingwei as the heirs of Sun Yat-sen or a confident gesture to suggest that the embattled government was still in charge of the nation.

This chapter, therefore, has not been merely an exploration of the ecological impact of trees but has also demonstrated how trees serve as a powerful platform that defines and sustains epochal political movements. As the urban-centric Nationalists became increasingly disconnected from the materiality of trees, the Communists came to heavily depend on the material and ecological benefits of trees for their very existence in rural regions. Trees have stood as living pillars of the Communists' political vision, with their roots intertwined with the very principles of resilience and endurance. The act of tree planting extends beyond an agricultural activity. Instead, it operates as a revolutionary endeavor that has sustained and nurtured the Communists, both materially and ideologically. In a challenging natural environment, it is the materiality of trees—their wood, roots, leaves, and fruits—that provides sustenance and shelter, becoming a testament to the Communists' ability not only to adapt but also to thrive. That materiality contributed to a forestry revolution after the Communists came to power in 1949 (see chapter 2).

In the tumultuous Republican era, the conflict between symbolic posturing and pragmatic implementation spawned various environmentalisms. This multiplicity highlights the fluid nature of environmental ideas, which cannot be reduced to a mere dissemination of knowledge from the United States to the Beiyang government and, later, to the Nationalists and finally to the Communists. It is by no means a linear relay race, nor can it be encapsulated under the blanket term *tree planting*. Instead, these ideas mutate, evolve, and regress as they interact with the distinct social and cultural contexts they encounter. Foresters, politicians, and educators all choose environmental ideas that foster their distinctive goals, resulting in the kaleidoscopic, mercurial, and flexible quality of modern Chinese environmentalisms. As will be seen in the next chapter, after the Communists came into power in 1949, environmentalism became increasingly contested even among environmental scientists themselves.

TWO

Selling the Forestry Revolution

The Rhetoric of Afforestation in Socialist China (1949–61)

Introduction

At the Second National People's Congress in 1960, Liu Wenhui, minister of the Forestry Department of the People's Republic of China (PRC), declared that his country was experiencing a "forestry revolution." Liu went on to say that he doubted that any other country was "pursuing a policy of forest restoration so broad, ordered, and planned" as was China. He compared China's forestry project to those of its Western counterparts, saying that "no senile capitalist country could compete against our socialist one."[1] This optimism provides a new perspective on the supposed irrationality of Mao-era practices and the destructive tendencies identified in early studies.[2]

A close examination of speeches and writings on the subject of trees and forestry in the early socialist period (1949–66) reveals a more nuanced story than the earlier reductive narrative of environmentalism in this era. To be sure, many Communist leaders did use militaristic rhetoric to pursue destructive, uncoordinated, and chaotic policies. The expressions "struggle against nature" and "war against nature" pervaded Maoist environmental newspapers and government documents. In national conferences and newspaper articles, scientists enthusiastically and overwhelmingly endorsed this so-called war. Nevertheless, their use of such expressions does not alter the fact that they were attempting to manage the environment from a scientific perspective. This is exactly where this chapter departs from previous scholarship in analyzing Maoist environmental discourse. Environmental scientists, as will be seen, considered the ecological impact of natural disasters and endeavored to address them in various ways. The story of tree planting presented here is intended to counterbalance the dominant scholarly emphasis on chaotic, irrational, and destructive rhetoric in this period. Unveiling the conflicting environmental notions among Mao-era environmental scientists will further our understanding of the contested environmentalisms of modern China—the key topic of this book. The scientists' competing notions not only reveal that environmental discourse during the Mao era contains significant elements that are scientific and constructive but also manifest how Chinese environmental science and technology made certain ways of grappling with the environment and certain patterns of environmental politics possible. Forestry science and hydrological science—though both committed to environmental management—show drastically disparate visions of preventing natural disasters, manifesting the ways in which tree-planting knowledge is integrated with other views on managing the environment.

Planting Socialist Nature

Foresters in the early socialist era actively promoted tree planting. This put them at odds with hydrologic engineers, who were focused on dam building. These debates reveal the dynamics and the contingency of China's environmental history. It is only through a closer examination of this history that the rationality of the actors involved can be fully appreciated.

Liang Xi (1883–1958), the chief minister in charge of forestry in the early PRC, is an excellent representative of the elite scientist (see figure 2.1). In his story, one can see both ruptures and continuities between the Republican era (1912–49) and the early PRC era with regard to their views of forestry. Though foresters during the Mao period rejected the legacy of the

Republican government, labeling it reactionary and counterrevolutionary, their endeavors actually continued the deeds of the earlier era rather than breaking with them, as the Communists claimed. Yet compared with that of their predecessors, the role of foresters in governmental institutions was reduced. The views of Liang Xi illuminate the evolution of forestry notions in early socialist China.

Liang's professional training and biography show that the origin of Maoist forestry is transnational in the sense that it derived from the capitalist bloc. Liang's views on Chinese forestry were heavily influenced by his experience studying abroad, especially in Germany. After studying forestry at Tokyo University in Japan in the 1910s, Liang pursued a graduate degree from 1923 to 1926 at the Forestry College at Dresden–Tharandt, which is one of the oldest academic forestry schools in the world and the birthplace of scientific forestry. There, Liang absorbed the German forestry conventions of that time, which strongly emphasized the centralized management of forests, making them easily supervised and harvested on a national scale.[3] German scientific forestry projected a utopian notion of easily manipulated

FIGURE 2.1 **The inscription at China's Northeast Forestry University in June 2019 reads, "Dressing up mountains and rivers with beauty; painting the national territory with red and green." It comes from Liang Xi's 1951 article "Forestry of new China" ("Xinzhongguo de linye") and stands as a goal for students in most forestry universities in China. Liang projected a utopian vision of foresters in China as artists. Photo by author.**

forests, but also a rigorous technical and commercial practice that could be readily disseminated. More importantly, compared to other German forestry schools that stressed forest utilization, Dresden–Tharandt adopted and promoted principles of forest conservation. These ideas greatly influenced and propelled Liang's career as a leading forester beholden to such ideas in China.[4]

The German emphasis on intense centralized forestry management had found a home in Republican China through its founder, Sun Yat-sen, who advocated forestry as a means of national salvation. As discussed in chapter 1, Sun's seminal political manifesto, "Three Principles of the People"—referring to the principles of nationalism, democracy, and the people's livelihood—published in 1924, articulated the linkage between forestry and people's livelihoods. He argued that forestry could improve the living conditions of the public and suggested that forestry held the key to preventing or ameliorating natural disasters. Probably for the first time in modern Chinese history, forestry was elevated to a central role in combating social and ecological problems.

In 1929, echoing Sun's manifesto, Liang published the article "The Problems of Livelihood and the Forest" ("Minshengwenti yu senlin") in the premier issue of *Forestry* (*Linxue*), the first journal to systematically elucidate the role of forestry in nation building. He argued for the centrality of forestry in all economic sectors, claiming that "clothes, food, living, and travel all depend on the forest. If a nation does not have a forest, the people will not survive."[5] Emphasizing the pervasiveness of the forest in daily life, he pointed out that shoes, socks, other clothing, ships, cars, and airplanes are all associated with trees. He noted, for example, that forest products formed an integral part of airplane production in Wisconsin in the United States when the industry was getting started in the 1920s.

The belief in the centrality of forestry science for national development in the pre-PRC period deeply shaped the forest policy of the PRC. In the inaugural 1950 issue of *Chinese Forestry* (*Zhongguo linye*), the leading forestry journal of the period, Liang anticipated the urgent need to promote forestry, especially tree planting (see figure 2.2). At the beginning of his seminal article, "Foreword to *Chinese Forestry*" ("*Zhongguo linye* fakanci"), he asserted that "nature treats China fairly, but the Chinese have rarely treated nature fairly."[6] He adopted the perspective of the *longue durée* (long view); ever since ancient times, the Chinese had exploited the forest randomly and carelessly, leaving only 50 percent of the land covered by forest. After castigating those ancestors who had wreaked havoc on the forests, Liang argued that the Chinese should henceforth show their gratitude to nature. Yet this gratitude, for Liang, was shown neither in Confucian humility before the natural world nor by a balanced and dynamic interlocked cycle between the five phases of ancient Chinese philosophy—fire, water, earth, metal, and wood.[7] Rather, it was expressed by the state's intervention in planting trees. Here, the German influence of centralized management looms large.

At the end of the article, Liang listed five major issues—forest management, forestry policy, forest protection, tree planting, and forest exploitation—that required development. He compared the forests to a feral child, "unclothed, unfed, unemployed, undisciplined, and uneducated," and argued that sound forestry practices could be compared to effective child rearing.[8] Forest management, if properly applied, would be like tracing the

FIGURE 2.2 **Cover of the inaugural 1950 issue of *Chinese Forestry* (*Zhongguo linye*). Source: *Zhongguo linye*.**

child's personality, history, knowledge background, and physical features. Forest policy would be akin to teaching the child self-discipline. Forest protection could be likened to feeding and clothing the child and making him stronger against bullies. Tree planting would be like "cleaning the blood, curing scabies, healing dysentery, restoring his hair, improving his skin and body, so that he is neither bald nor has scars, and he does not feel shame in front of the public."[9] By using the dreadful situation of a wild child to illustrate the horrible situation of China's forests, Liang tried to mobilize people to plant more trees. In regard to forest utilization, Liang argued that it is similar to guiding the wild child to participate in production exercises contributing to society. However, this utilization should be limited. He thus demonstrated his faith in managing natural resources for China's long-term development by means of forestry science. Problems arose, however, from scientists in other disciplines who also aspired to remedy China's natural ills.

One of Liang's most prominent endeavors in promoting forestry in the PRC stemmed from his firm belief that forestry was pivotal in solving the chronic problems related to managing the Yellow River without resorting to expensive hydroelectric dams. The Yellow River has been associated with trauma and regime legitimacy since ancient times. For centuries, floods of the Yellow River have drastically affected settlement patterns, agricultural production, and social stability. Ling Zhang points to the economic damage that the river caused in medieval China, and Micah S. Muscolino discusses the moral conundrum that the Republican government encountered after it destroyed the dikes to create a flood to stop the Japanese invasion.[10] In both cases, the Yellow River called into question the legitimacy of governments.

Compared with other rivers of the world, the Yellow River carries the most silt. Ruth Mostern highlights that from the ninth to the twelfth century, the Yellow River basin experienced a drastic increase in sediment discharge primarily owing to population growth, military defense, and the burgeoning timber industry. This trend continued into the eighteenth and nineteenth centuries, with ongoing population expansion and timber commercialization significantly contributing to persistent soil erosion in the region.[11] David Pietz shows that in the Republican era, the Yellow River's high load of silt frustrated European engineers, who realized that their hydrological theories could not be applied to China.[12] In order to address these issues, hydrologists borrowed from the foresters in the Republican era. In 1940, Ling Daoyang, the leading forester of that time and Liang's predecessor (examined in chapter 1), was appointed as the vice director of the Yellow River Commission.[13] Ling promoted tree planting for the conservation of water and soil and considered planting trees a valid approach to solving the problems of the Yellow River.[14]

Liang Xi continued Ling's approach and further privileged forestry over hydrology in managing the river's natural problems. In the inaugural 1950 issue of *Chinese Forestry*, Liang contended that sand constituted 50 percent of the Yellow River's flow and that planting trees could effectively control flooding on the river. Moreover, planting trees could cause part or even all of the Yellow River to become transparent; collaborating with another forester, he imagined that "the Yellow River [could] become a Green River."[15] In 1950, when he addressed the Northwest Agricultural Technology Conference, Liang reiterated his conviction that tree planting could reduce riverbed silt. In neither instance did he mention dam building.[16]

The Water Resources Department, however, had a different approach to solving the challenges posed by the Yellow River. The leading hydrologist in Maoist-era China and the chair of the Technical Committee of the Water Resources Department, Zhang Hanying, articulated his opinion as a hydrologist in the *People's Daily*.[17] Describing water-flow changes and the associated natural disasters that had plagued China for centuries, he suggested that the fundamental solution to the problems of the Yellow River was to manage the water flows. Educated at the University of Illinois and at Cornell University, Zhang was convinced of the effectiveness of modern engineering—that is, building dams—to solve those problems. According to Zhang, the sand in the water and the flooding of the Yellow River came from the same geographical areas. To solve the silt problem, in essence, one had to grapple with the water of those geographical regions directly. Specifically, it was the middle part of the Yellow River that Zhang identified as the crux of the Yellow River's problems. According to Zhang, building a dam on that part of the river would remove the silt, regulate the river's flow, facilitate irrigation and shipping, and generate electricity—all at the same time. For Zhang, the water would become a resource after the dam was built on the middle region of the river, which would help to ensure sufficient water resources for the lower region in the dry season.[18] Fu Zuoyi (1895–1974), the minister of the Water Resources Department, supported this initiative and agreed to build a dam on the Yellow River in 1950.[19] In 1951, projects began to move the Yellow River's water to the Wei River, a 283-kilometer-long tributary of the Hai River. An article in the *People's Daily* promised that thanks to the building of the dam, water problems would become water resources.[20]

Liang Xi, however, took issue with the actions of the Water Resources Department and expressed disagreement with the hydrologists in regard to the scale of the environmental problems. On September 22, 1950, at a meeting of experts in agriculture and forestry in northwestern China, Liang focused on the deeper causes of the Yellow River's silt. Unlike Zhang, who identified a small area as responsible for flooding, Liang pointed to a much larger region as the cause of the problem. He highlighted the importance of forest conservation on the Loess Plateau in Shaanxi and Gansu Provinces, which comprise the vast middle part of the river. The conflicting views reveal the contrasting approaches of the hydrologists and the foresters. For the former, the water was the object to be managed rather than the sand. The foresters took the investigation one step further by exploring the cause of the sand in the Yellow River. They pointed to the interaction between soil and water, reflecting a broader vision of nature. In other words, the hydrologists saw the natural world as relatively *static* by narrowing their focus to the water flows, while the foresters held a more *dynamic* view of water, trees, and soil interacting as one whole.

The friction between the two groups entered a new phase when the minister of water resources management and the minister of forestry traded speeches at a national conference in 1951. During the third session of the First National Committee of the Chinese People's Political Consultative Conference, Fu Zuoyi suggested that the fundamental way to solve the silt problem was either by strengthening the fragile embankment or by building a dam. Although he granted that water and soil conservation could be advanced by planting trees, he disparaged tree planting as a long-term project that would require labor on a large geographical scale.

In his speech, Liang reminded his audience of the danger of soil erosion.[21] He declared that planting trees was fundamental as well as urgent:

> If we cannot protect the soil on the surface of the mountain, on the upper river, as Fu has said, "the land is washed and becomes more and more barren." On the lower region, as Fu has said, "the riverbed is elevated because of sand and soil, becoming a ground river." Not only the riverbed but also any dam will be slowly blocked by soil and water. Jinghui Canal and Luohui Canal in Shaanxi were supposed to be used for irrigation. However, last fall, some places did not dare to open the sluices to release water because they were afraid that they would end up releasing mud. Releasing mud poses problems instead of contributing to the farmland.[22]

Superficially, he agreed with Fu by citing his speech. Nevertheless, Liang elevated the urgency of tree planting by saying that building a dam would be futile if there were not sufficient trees to guard against soil erosion. Liang projected a pessimistic or even apocalyptic vision of the future if tree planting was not addressed immediately.

Liang appeared to have lost this discursive battle since dams became the main means to solving the problems of the Yellow River. The Soviet Union, which had supported many hydraulic programs in China between 1949 and 1952, played a crucial role. Soviet experts offered modern hydrological expertise to build dams on the Huai River during this period, and with Soviet guidance, socialist China constructed a dam that proved effective in managing the Huai River, staving off famine and generating power. The success with the Huai River dam convinced the Chinese Communist Party (CCP) that another should be built on the Yellow River, and in 1954, China invited Soviet engineers to investigate the possibility of building dams there.[23]

These efforts are associated with what David Pietz has called a "technological complex" in the socialist period. According to Pietz, the complex "connotes the entire range of goals, actors, material, institutions, and tools organized to support a single project or a set of projects."[24] Driven by this technological complex, a magnificent dam functions as a monument to the immediate and tangible accomplishments of China. Liang, however, cast doubt upon these eye-catching dam-building initiatives and, somewhat ironically, strove to combat the Soviet Union's dam-building influence discursively by using his knowledge of the Soviet Union.[25] In 1954, in his significant monograph *The Role of Forests in National Economic Construction (Senlin zai guojia jingjijianshe zhong de zuoyong)*, Liang highlighted the accomplishments of the Soviet Union in grappling with drought and flood by tree planting.[26] By citing Stalin's Great Plan for the Transformation of Nature, Liang conferred legitimacy on China's tree-planting projects.[27]

In fact, Soviet forestry development was undergoing rapid changes at precisely the time of Liang's monograph. Launched in 1948 as the world's largest ecological engineering project theretofore, the Great Plan for the Transformation of Nature was basically an afforestation plan, which reorganized Soviet foresters for hydrological reasons. The project was driven by the conviction that a natural environment without forests would be inherently unstable. Liang was smart to use the Soviet plan as evidence to promote forestry in China. Yet this plan was hardly as positive as Liang described. Stephen Brain shows the tension of the Great Plan for the Transformation of Nature as between the "technocrats . . . [who] promoted an approach to forestry that acknowledged local variation, natural limits, and the importance of experimental results, with special scientific training required to determine the best outcome, [and] the prometheans [who] advanced the notion that forests could be made to conform to the human will and relied upon ideological claims to back up their beliefs."[28] Although the plan was ambitious in its scale from 1948 to 1953, it was halted after Stalin's death in the latter year. The Forestry Department was liquidated, causing forestry conservation to decline sharply from 1953 onward.[29] Liang chose not to mention these recent developments.

Liang's one-sided description of Soviet forestry practices was probably intentional. Because Liang, as China's forestry minister, had the privilege of receiving the latest information about Soviet forestry, one may assume that by 1954, he was aware of this sea change and of the competing ideologies of forestry that had existed for years in the Soviet Union. Nevertheless, he still argued that Soviet tree planting was widespread and highly scientific. He cited Georgii Morozov (1867–1920), a forerunner of the Soviet Union's technocrats and the father of modern Russian forestry, who argued that forests contributed to precipitation because the air over them was cooler than over unforested land. He also mentioned the experience of Soviet aviators who claimed that the air pressure is lower over wooded land, concluding that rain is likelier in forest regions because of air convection. In addition, he noted that the Moscow Agriculture Academy had proved that the precipitation of forested regions is 17.4 percent higher than those without trees.[30]

Armed with the rhetorical weapon of Soviet forestry, Liang systematically examined and justified the scientific foundation of tree planting for the first time and further suggested afforestation as a technique that could be applied to the entire region of the Yellow River. Though he now, as before, claimed that Shaanxi and Gansu Provinces (the regions around the middle of the Yellow River) suffered drought exactly because of the cutting of trees, he also argued that planting trees would be beneficial for the upper region of the Yellow River. The upper region, functioning as the headlands of the river, was therefore pivotal to the entire river. In other words, Liang was using a scientific principle taken from the Soviet Union to justify his initiative to expand the geographical scope of tree planting on the Yellow River.

In addition to his adaptation of Soviet practices to advance China's tree planting at home, Liang also promoted China's tree planting on an international stage. The image of rational and constructive afforestation coincided exactly with the image that China wanted to promote internationally, especially in the aftermath of the war on the Korean Peninsula. In 1954, Communist China had the opportunity to earn the respect of the international community by describing its success in afforestation during the fourth meeting of the World Forestry Congress held in Dehradun, India.[31] Notably, unlike the previous three conferences, which stressed the productive use of forest products, the 1954 conference focused on the protective functions of trees—namely, soil and water conservation. In the same period in East Asia, Japan had just begun its program of arboreal rejuvenation.[32] In Korea, Paul Zehngraff, an American forester who championed a United Nations–sponsored working group survey on South Korean forests in early 1950, advocated afforestation against soil erosion on Korea's bare ground.[33] Though it cannot be determined whether the experience of Japan and South Korea influenced socialist China, Chinese foresters were indeed aware of this recent trend when they attended the conference. Liang selected and sent three Chinese films about the protective functions of forests to be presented at the gathering.[34] Here, one can see a particular dynamic between socialist China and the outside world: on the one hand, the most recent international trend in forestry helped direct China's forest policy; on the other hand, China wanted to use its achievement on the protective functions of trees to present a positive image at a time when it was isolated by the capitalist bloc during the Cold War.

While Liang endeavored to advance forestry at home and abroad, he showed great flexibility in integrating his ideas with those of hydrologists after the dam construction on the Yellow River had begun. Instead of mutual disregard, foresters and hydrologists began to learn from each other and drafted a reconciled policy. Probably realizing that he could not prevent the construction of the dams, Liang shifted his strategy to emphasize the benefits of forests for the dam rather than entering a debate on whether trees or dams were more vital to solve the water problems of the Yellow River. He mentioned that in July 1936, the Soviet Union had made a law to conserve water by planting trees, and he hailed this law as the first in world history to adjust water on the ground's surface by forests. Liang's introduction of the Soviet law that links forests with water aimed at justifying Liang's afforestation in China.[35] He suggested that tree planting can prevent soil erosion, which negatively affected the operation of the dams on the Yellow River. Liang's introduction and adaptation of Soviet tree-planting strategies as well as his repurposed arguments with regard to the trees' benefits for the dam show that his relationship with the hydrologists slipped from antagonistic criticism to reconciliation.

Just as Liang accepted the opinions of his colleagues from the Water Resources Department and no longer rejected building dams, hydrologists compromised by adding afforestation to their hydrological plans. Vice Premier Deng Zihui (1896–1972) highlighted tree planting in grappling with the Yellow River's problems during the second session of the First National People's Congress in July 1955. He incorporated Liang's suggestion about the role of trees in reducing silt, mentioning that "the fundamental reason for the flood in the lower region of the Yellow River is soil erosion in Gansu, Shaanxi, and Shanxi Provinces."[36] At the same time, hydrologist Zhang Hanying also echoed Deng's remarks regarding planting trees on the Yellow River. He wrote that the silt accumulated at the Sanmenxia Dam would be halved after the first round of tree planting was finished.[37] This was a huge contrast to Zhang's former approach to managing rivers. In his earlier study of the Yellow River during the Republican era, Zhang argued that the silt from the Yellow River caused the river to change its course. He did not, however, investigate the root cause of the high silt and sediment concentrations.[38] As mentioned earlier, Zhang also failed to take tree planting into his 1949 account. With this in mind, Zhang's inclusion of tree planting among the solutions to the Yellow River's various challenges indeed demonstrates the influence of foresters.

The friction between hydrologists and foresters reveals a nuanced but nonetheless significant aspect of Chinese environmental management. Instead of considering environmental policy as a unified whole, two disparate voices of environmental rhetoric competed and negotiated with each other. Citing the experience of the Soviet Union, Liang Xi stressed tree planting in environmental stewardship at a time when Chinese hydrologists—also under the influence of the Soviet Union—promoted building dams. While hydrologists tended to focus on more immediate and small-scale problems, foresters preferred to identify and solve the fundamental problems of water disasters and soil erosion and stress the interaction between water and soil in a larger ecological system. Foresters considered environmental problems on a broader conceptual scale and a more universal geographical scope and in terms of a more dynamic interaction between various natural agents.

On a deeper level, the friction in socialist China between Chinese forestry science and hydrological science made certain kinds of environments and certain patterns of environmental politics possible, far beyond a simplistic portrayal of socialist science and technology either as powerful tools of progress or destructive agents of decline.[39] Neither science prevailed. Rather, they penetrated one another, forming a dialogical relationship. In this process, socialist China's environmental policy derived not only from the Soviet Union but also from Germany and the United States. Environmental rhetoric and policy in Mao-era China were therefore more resilient, polyphonic, and transnational than scholars have acknowledged.

Mao's Greening the Motherland Campaign

Inspired by Liang's blueprint for solving the Yellow River's problems, Mao Zedong integrated afforestation into his social-engineering and mass-mobilization plans from the late 1950s onward. Thanks to Liang's promotion of the ecological roles of forests in the Yellow River Valley, Mao emphasized the role of trees in that region and beyond. If Liang Xi, as a forester, mainly promoted tree planting from a scientific perspective, Mao elevated it to a political imperative—a national campaign.

The development of the campaign should be attributed not so much to Mao as to his forestry minister, Liang Xi. His commitment to the Yellow River was firm and unapologetic and eventually influenced Mao. When foresters and hydrologists spoke at the national conference, Mao was present, and this probably elevated the importance of tree planting in his mind. In March 1956, when Mao responded to the youth from the Yellow River Valley, he encouraged them to "green the motherland" (*lühua zuguo*). Indeed, this constituted a dramatic change after Mao addressed the Yellow River's problems as part of his national vision. Soon, the Greening the Motherland initiative expanded throughout the country and became a nationwide campaign (see figure 2.3). During the same year, Mao's ambition to plant trees nationwide was included in the draft of the National Outline for Agricultural Development (1956–67), which aspired to plant trees along all roads, in all residential areas, near all bodies of water, and on all barren land nationwide within twelve years.[40] Such emphasis on tree planting for Mao was by no means accidental, which is clear if we consider his earlier experience as well as the aforementioned influence of the foresters.

In a geographical sense, Mao's Greening the Motherland Campaign in the 1950s, launched in the Yellow River region, was strongly linked to his struggle for survival in the same area during the 1930s and 1940s, as discussed in chapter 1. This association is further highlighted by the grueling journey of the Red Army's Long March in the mid 1930s. The challenging journey took the Red Army to Shaanxi, where barren lands and soil erosion threatened the Yellow River with floods, posing a risk to Mao's precarious government. While Mao won the Civil War in 1949, his desire for survival did not fade. The escalating tensions in the Korean Peninsula in the early

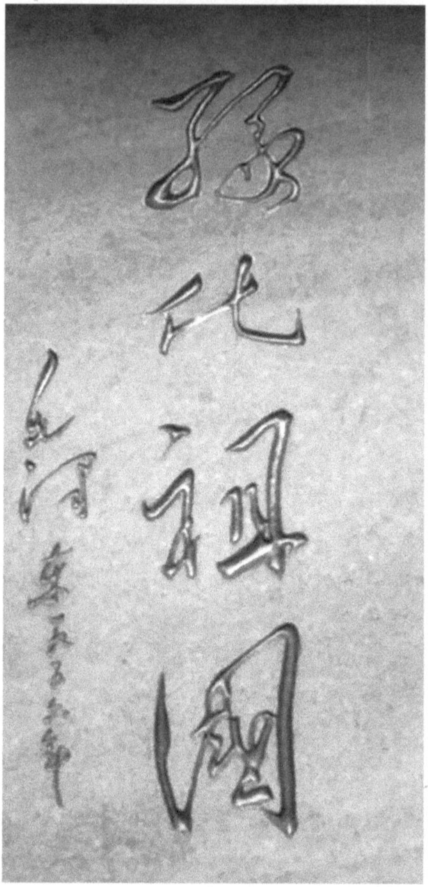

FIGURE 2.3 **Photograph of Mao's epitaph, "Greening the motherland," June 2018. Photo by author.**

1950s and the continuing threat from the Nationalist government even after its retreat to Taiwan in 1949 compelled Mao to continue to put emphasis on the survival of his regime. His Greening the Motherland Campaign of the 1950s and 1960s was launched from Yan'an and was not so much an ecological initiative to improve the Yellow River Valley as a continuation of his desire to guarantee the survival of his revolution.

The Great Leap Forward (1958–61) accelerated the momentum of tree planting, at least in a rhetorical sense.[41] In 1958, the *People's Daily* reported that the youth in Shanxi Province claimed that they had planted one hundred million trees along the Yellow River in a month.[42] In 1959, the newspaper reported that tree planting appeared to have expanded nationwide and that provinces were competing against each other to plant the most trees.[43] Though some scholars cast doubt on the actual results of the tree planting, it is undeniable that during the Great Leap Forward, afforestation was promoted as a national initiative.

The expansion of tree planting to a national campaign meant that knowledge of how to do it would have to be widely disseminated and could no longer be restricted to scientific elites. Writers were instrumental in this process: an examination of the literary writings published during the Great Leap Forward shows how authors facilitated the transmission of the requisite knowledge from the experts to the grass roots. One such work was the pamphlet *Barren Hills Blossom with Fresh Flowers* (*Kushan kai xianhua*), published in 1960 by the Chinese Forestry Press (see figure 2.4). The work displays the considerable rhetorical skill that the Communist government employed to inculcate knowledge of tree planting in the common people, along with populist discourse on gender equality and social liberation.[44] The title story, originally published in the *People's Daily* on January 10, 1960, depicts girls and elderly women planting trees on a barren mountain in a strong wind. The text focuses not only on their upbeat spirit but also on their professional skills. Understanding the methods of conservation of water and soil allowed sixteen-year-old Zhang Guihua to grapple with the desert conditions: "Even though I am young," she says, "I can still manage the mountain."[45] The mastery of tree-planting skills by young females empowered them in a traditionally male-dominated society, manifesting the socialist endeavor to advance gender equality.

While the first story shows that women's acquisition of knowledge of tree planting is an important national need, the second concerns the need to transmit this knowledge to a wider range of women. In "10,000 *Mu* [667 hectares] of 'Women-Planted Forest'" ("Wanmu 'sanbalin'"), the seedlings the women planted were not suited to survive.[46] Some were buried too deep in the soil; others, not deep enough. The men begin to question the ability of the women, saying that they would rather let the seedlings die on the ground than let them be planted and die in the dirt. The accusation of incompetence engendered shame and mobilized the women not only to work harder but also to learn the specific skills of tree planting. One

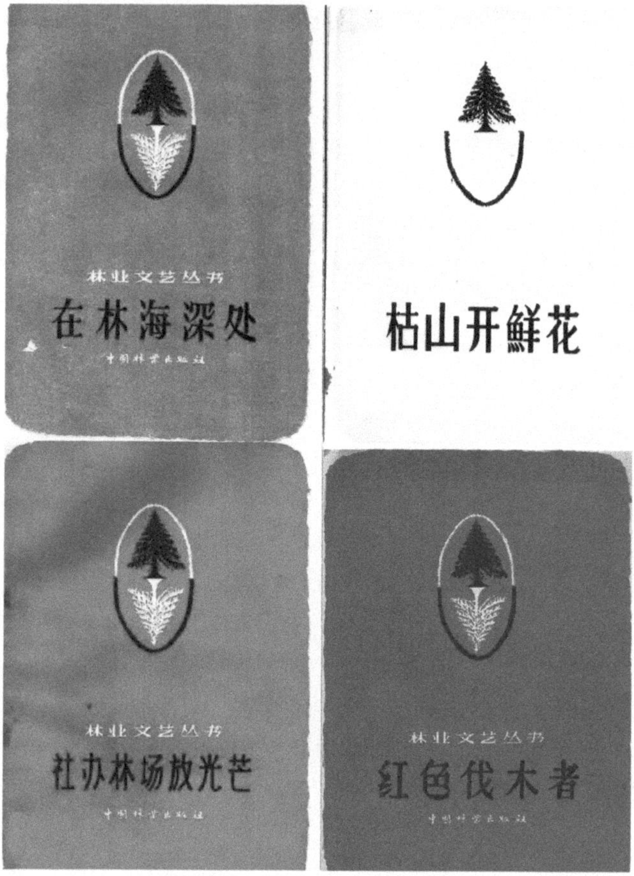

FIGURE 2.4 Four pamphlets in the Forestry Literature Series: *top left, In the Depths of the Forests* (*Zai linhai shenchu*) lauds areas with enticing natural resources, especially northeastern China; *top right, Barren Hills Blossom with Fresh Flowers* (*Kushan kai xianhua*) chronicles planting trees in a harsh natural environment; *bottom left, The People's Commune Forest Farm Shines Brightly* (*Sheban linchang fang guangman*) is a collection of poetry eulogizing the planting of trees; *bottom right, Red Loggers* (*Hongse famuzhe*) collects three news reports about model workers who cut down trees to contribute to socialist construction. Sources: *Zai linhai shenchu*; *Kushan kai xianhua*; *Sheban linchang fang guangman*; *Hongse famuzhe*.

woman arranged to learn the skills from her son.[47] She then began to transmit this knowledge to her peers. It proved not to be an easy task, however. The story implies that acquiring the skills of tree planting is a process of trial and error. After half a month, more and more women began to master these skills.[48] Finally, their skills empowered them and raised them to the same level as the men. In conveying the rhetoric of gender equality, the story, on the one hand, masks the elitist discourse of tree planting as an activity requiring professional training. On the other hand, it suggests that tree planting demonstrates the growing power and ability of women—an agenda that Mao's government eagerly promoted.

The Greening the Motherland Campaign urged the younger generation to get involved in national construction. Captioned "Youngsters! Use your own two hands to make our great motherland greener," a 1956 campaign poster shows Chinese state power calling upon the youth to participate in tree-planting initiatives by evoking patriotic sentiments (see figure 2.5). In the background, a row of trees has begun to cover the barren mountains. The caption suggests that the closer they are to nature by planting trees on Chinese soil, the deeper and more intimate is their connection to the Chinese

FIGURE 2.5 **Poster captioned "Youngsters! Use your own two hands to make our great motherland greener," 1956. Courtesy of ChinesePosters.net, IISH collection.**

nation. Once linked to the nation, they become an integral part of the national agenda advanced by the Chinese state power. While the boy holds a seedling on his shoulder, the girl carries a shovel and summons other youth to join them. The poster not only indicates the equal status of women and men but also reflects the process of molding the youth into mature citizens who are concerned with their natural surroundings through collective actions. While planting trees, they also learn to cooperate and work as a team under the guidance of the socialist government, which transforms them from observers into active participants. Consequently, this experience fosters a close-knit camaraderie and a forward-looking viewpoint.

While tree planting was contributing to the toughness, resilience, and growth of youth into adulthood, local news reports integrated forestry with other agendas, such as agricultural production. In Xingtai, Hebei Province, the youth shouldered the responsibility of planting trees on rocky soil and in mountains. A Forestry Department–endorsed report says that "they marched into nature with perseverance and heroic motivation."[49] Another report claims that the youth in Baotou, Inner Mongolia, were resolutely determined to achieve their tree-planting goals within seven years and rose to the challenge.[50] In some places, young people planted trees to fight soil erosion for reservoir safety and agricultural production; in other places, they planted fruit trees. Tree planting not only molded youth into adulthood but was internalized and integrated with other prominent social and political goals for nation building in the Mao era. The relatively marginal tree-planting initiatives survived and thrived.

Conclusion

This chapter has explored the disparate voices in environmental management during the Mao era. The friction between hydrologists and foresters has showed the diversity and multiple possibilities of environmental stewardship in socialist China. Thanks to the endeavors of foresters, tree planting became an essential part of nation building, especially after Mao's launching of the Greening the Motherland Campaign.

The chapter has demonstrated not only that the rhetoric of forestry and of environmentalism in general during the Mao era had significant scientific and rational aspects but also, and more importantly, that there were disagreements among environmental scientists themselves. The source of environmental knowledge, both in forestry and in hydrology, was Western science. The friction between foresters and hydrologists in China did not result from political or factional clashes but from conflicting scientific principles. In spite of their differences, both sides addressed ecological problems in a scientific and objective manner. Apart from the scientists, Mao himself was also sober and sensible in launching the Greening the Motherland Campaign to tackle ecological difficulties. In a broader geopolitical context, the Cold War tensions that threatened to annihilate Mao's government further pushed him to consider the immediate ecological challenges that could lead to drought and famine. Capitalizing on Mao's campaign, foresters skillfully employed socialist political language to promote tree planting so that forestry knowledge would not be accused of being elitist. In other words, foresters were well aware of their risk of being marginalized or isolated, and they endeavored to stay away from this trap. These rational aspects of environmental ideas and policies played an important role in nation building during the Mao era.

Moreover, it has been shown that China's emphasis on tree planting as a means of nation building should be understood in a global context, which demonstrates the flow of conservationist notions from the outside world to China's villages and from scientists to the general public. Germany influenced Liang Xi during his formative period regarding the role of forestry and its management methods. The Soviet Union's experience of afforestation later justified Liang's efforts concerning tree planting and functioned as a rhetorical weapon against China's leading hydrologist, Zhang Hanying, who had been educated in the United States. More importantly, socialist China's Soviet-influenced rhetoric on tree planting was an integral part of a global turn in the 1950s to the protective function of forestry. Rather than passively follow their foreign colleagues, China's foresters actively engaged with them by sending films of tree planting to international conferences. In so doing, they aspired to present a positive image of China during its isolation in the Cold War era. As a result, while Chinese foresters appropriated transnational knowledge, tree planting, in turn, became a cogent diplomatic calling card for China.

Mao's nationwide campaign was part of the transmission of the global knowledge of tree planting to China's local level. To instill this internationally acquired knowledge in people in outlying regions, foresters skillfully integrated elite discourse of tree-planting expertise with the dominant socialist revolutionary discourse on topics such as women's liberation and youth mobilization. Tree-planting discourse in China was thus inherently transnational yet local in practical terms. These two features propelled foresters to influence environmental thinking in the citizens, including women and youth, of the country with the then-largest population in the world. Forestry was no longer merely a tree-planting project when it was integrated with other national agendas and gained a foothold in public discourse. But when the Greening the Motherland Campaign promoted by the state power arrived in the borderland, resistance arose from local ethnic groups. This reaction is depicted in a Mao-era film to be explored in the beginning of the next chapter.

THREE

Marching into the Desert

Planting Ethnic Borderlands in Mao-Era Ecocinema

Introduction: Sinification by Greening

While tree planting can be used to popularize conservationist ideas, it can also go astray and become part of an insidious agenda when it comes to the borderlands. When Mao Zedong extended the Greening the Motherland Campaign to ethnic minorities in the border areas, fierce conflicts arose between the Han Chinese and other ethnic groups. These tensions between Mao-era conservationist efforts and environmentalism are visible in cinematic representations of afforestation in the 1960s and 1970s.[1] If the official declarations found in the CCP mouthpiece the *People's Daily*, explored in the previous chapter, provide citizens with the reasons to plant trees, the ecocinemas examined in this chapter dramatize the conflicts and illustrate the final goals of afforestation in the borderland. The narrative of the borderland is by no means marginal to the central issues of Chinese environmentalism.[2] Understanding what qualifies as *Chinese* environmentalism requires visiting China's most "foreign" areas because the most exotic regions vividly demonstrate the desires and goals of state power.

Considering China's tree planting in the borderland also offers a unique approach to considering the interaction between colonialism and environmentalism. In Western history, European colonizers disrupted local ecosystems in North America by importing plants and animals from home in what Alfred Crosby calls "ecological imperialism."[3] For white settlers during the eighteenth and nineteenth centuries, clearing the land to build new living spaces and increasing food production were deemed "natural" processes for "improvement." The British Empire's colonization of India through the British East India Company also caused severe deforestation.[4] Although European colonialism now attracts widespread scholarly attention, colonialism is not, of course, a European invention. Recent scholarship in Sinophone studies and New Qing History explores China's imperial ambitions throughout the Eurasian continent, beginning in the Qing Empire. Since the seventeenth century, driven largely by demand for natural resources, China expanded its territory as it marched toward the west.[5] This expansion had enduring consequences in environmental alteration in the twentieth century. The newly established Communist regime, dominated by Han Chinese, also expanded its rule to borderland regions and peoples and reshaped Indigenous lands and environments. It is noteworthy that, unlike most modern Western empires that deforested their colonies, Mao-era China tried instead to afforest borderlands in order to Sinicize the ethnic minorities who lived there.

While trees, as a vehicle of state power, acquired "universal" value for the Han Chinese in Mao-era China, the local environmental knowledge of ethnic groups fell gradually into oblivion. Traditional approaches to grappling with the harsh (mostly desert) environment inherited from ancestors were channeled into a uniform method: tree planting. Admittedly, greening the wilderness improves the productivity of the land and feeds more people, including the Han migrants. An inconvenient truth, however, is that framing the tree-planting campaign as a sweeping national crusade downplayed and even erased the cultural understandings of nature held by the Indigenous population. When the Han Chinese imposed their environmental attitudes on ethnic minorities through tree planting in an inhospitable land, the less visible and more enduring impacts that afforestation entails were shifting land ownership to the nation-state. In other words, the Communists' beliefs regarding the "universal" benefits of planting trees to conquer the hostile world were inevitably entangled not only with the project of transforming the ethnic minorities' Indigenous environmental cultures and practices but also with gaining their allegiance to the nation-state.

In *Seeing like a State,* James C. Scott explains how the authoritarian state neglects local traditions and variations by setting up standardized formulas. These uniform prescriptions often lead to failure to implement national schemes on the local level.[6] Likewise, anthropologist Anna Tsing argues that "in the matrix of colonialism, universal reason became the mark of temporally dynamic and spatially expansive forms of knowledge and power. Universal reason, of course, was best articulated by the colonizers."[7] For them, local and situated knowledge of ethnic minorities does not matter in nation building and colonial expansion. Accordingly, planting trees as a universally progressive crusade without considering local environmental traditions becomes inherently regressive and oppressive through the dangerous liaison of greening and Sinification.

This chapter will first explore the scientific discourse concerning the conquest of the desert put out by the Communist government. Portraying the desert as an enemy to be defeated and exploited, these treatises contributed to an understanding of tree planting as a panacea that could be widely applied. While the scientists' rhetoric promoted the universal benefits of tree planting, films reveal the process by which the purportedly universal knowledge is spread and received at the local levels. It then examines a feature film, *Spring in the Desert* (*Shamo de chuntian*; 1975), which illustrates how resistance from the local ethnic population emerges when universal values are at odds with local ones. In *Spring in the Desert,* Narenhua, a Han Chinese–influenced Mongolian woman who believes that tree planting will help fight desertification, clashes with her fellow Mongolians, led by Damulin, who prefer to migrate to escape the growing number of sandstorms.[8] While, on the surface, the film depicts a conflict between two Mongolians, at a deeper level, it portrays the conflict between the Han Chinese settled lifestyle and the nomadic way of life embraced by the Mongolians. While *Spring in the Desert* demonstrates this tension in an overt manner, the documentary film *The Army's Reclamation and Battle Song* (*Junken zhange*; 1965) provides a less obvious narrative concerning Han attempts to shape the environment in Mao-era Xinjiang that overlooks the voices of the Indigenous population entirely.[9] In *The Army's Reclamation and Battle Song,* planting trees not only instills environmental ideas into an ethnic group, as in *Spring in the Desert,* but also legitimizes government control in ethnic regions under the veneer of conservation.

Examining the scientific discourse regarding afforestation in Mao-era China, as well as ecocinema dealing with the same topic, illuminates the conflicting attitudes of ethnic minorities and Han Chinese in the Mao era and further exposes an inherent tension between universal and localized environmentalism. By regarding tree planting as a universally positive initiative, the Han-dominated Communist government eliminates other ways of seeing nature—in particular, the ways of ethnic minorities. Tree planting imposes values on these minorities by devaluing the nomadic lifestyle, acculturating the minorities in Inner Mongolia and Xinjiang into new conceptions of nature, and, finally, driving them to pay homage, either consciously or subconsciously, to the newly established nation-state rather than to divine nature as they had traditionally done. In addition, the success of the greening campaign is presented as evidence of the Communist Party's superiority over its predecessors in mobilizing ethnic minorities to defeat the harsh natural environment. The party is legitimized by portraying itself as an emancipator of the oppressed classes within the ethnic minorities and as serving the project of socialist nation building. Depictions of ethnic minorities in Mao-era ecocinema thus serve as a prism through which one can perceive the dangerous liaison of acculturation and environmental changes in modern China.

"Marching into the Desert": Environmental Science Took Root in the Local

The project of planting trees in the Chinese borderland needs to be understood within the context of the global transmission of modern environmental science. This global narrative becomes more visible when one traces the genealogy of early Mao-era forestry. As was seen in chapter 2, the forestry minister in the early Mao era, Liang Xi, was educated in Japan and Germany. Liang strove to integrate tree planting into the Chinese national project, eventually resulting in Mao's launching of the Greening the Motherland Campaign. The idea of greening was later extended into the desert in the borderland. A brief examination of the writings of the renowned meteorologist and vice president of the Chinese Academy of Sciences (Zhongguokexueyuan) Harvard-trained Zhu Kezhen suggests how scientific ideas can serve as a powerful and sweeping universalist discourse that may ultimately turn hegemonic and oppressive.

Zhu took up Mao's imperative to plant trees and applied it to the desert. In 1959, he published the seminal article "Transforming the Desert Is Our Historical Mission" ("Gaizao shamo shi womende lishirenwu") in the *People's Daily*.[10] Zhu utilized multiple disciplines—including meteorology, forestry, hydrology, and agriculture—in his battle against desertification. This vision of a united front led him to bring together the expertise of the various government departments in charge of environmental management to formulate an integrated plan.

Zhu's project of reengineering the desert helped relieve the tension between hydrologists and foresters and sustained the momentum of Mao's Greening the Motherland Campaign, examined in the previous chapter. In October 1958, a national conference dedicated to finding ways to fight desertification was held in Inner Mongolia. It urged residents of northwestern China to "combine transformation and utilization of the desert to achieve comprehensive greening."[11] The conference attempted to transform desert areas into centers of animal husbandry and forestry, improve the soil, and change the climate.

Zhu's ideas on how to address deforestation were twofold. First, he contended that socialist China was more competent in grappling with desertification than either its predecessor governments or Western powers. He claimed that capitalist society had not succeeded in transforming desert into farmland; instead, it had only caused more desertification. In the United States, Zhu pointed out, deforestation in Kansas and Nebraska had caused millions to lose their farms in the Dust Bowl's droughts and sandstorms of the 1930s. He suggested that French colonizers in Africa had likewise caused desert expansion because of their deforestation practices. Zhu asserted that Western countries did not really want to help African countries control desertification: while US president Dwight D. Eisenhower expressed his eagerness to transform the Sahara Desert in Africa, what he really cared about was oil and other natural resources in the region. In contrast to these failed attempts by Western powers, Zhu argued that local governments in Xinjiang, Inner Mongolia, Shaanxi, and other northern provinces had succeeded in mobilizing peasants to plant trees and thereby control the desert. By contrasting Western countries' failure in the battle against desertification with China's success, Zhu advanced his scientific agenda of fighting against desertification in China.

The second thread of Zhu's efforts set the struggle against desertification as a task facing the entire human race by depicting deforestation and desertification from the perspective of the *longue durée*.[12] In 1961, targeting the masses, Zhu published "Marching into the Desert" ("Xiangshamo jinjun") in the *People's Daily*, in which he claimed that the civilizations of ancient Egypt, Babylon, and Greece had all been devastated by desertification. In addition to once again blaming capitalism for its role in creating the Dust Bowl in the United States, the article also examined China's own history. Owing to oppression by the emperor and landlords, farmers had been migrating to the provinces of Shanxi and Shaanxi since the late eighteenth century. But they did not irrigate the land they found there, causing the soil to lose its fertility and eventually giving rise to desert.

As Zhu integrated scientific and military discourses to promote afforestation of the desert, he regarded scientific methods as a powerful universal weapon that could be widely applied. Zhu argued that the desert used sand and wind as weapons.[13] According to Zhu, the desert launched its "attack" either in the form of guerrilla or positional warfare. Wind piled up sand dunes and advanced slowly. The dunes' height generally ranged from four to fifty meters, with some reaching as high as one hundred meters. Although the dunes moved slowly, they posed a devastating, even apocalyptic, threat, leaving forests destroyed, gardens buried, and cities reduced to ruins in their wake. Zhu argued that planting trees was one crucial weapon to fight against the malicious desert. As evidence, he pointed to Inner Mongolia and Xinjiang: thanks to sufficient water, trees now thrived; these trees, in turn, enabled railroads in Inner Mongolia to run smoothly by keeping the sand off the tracks.

At the local level, Zhu's science-based efforts at fighting desertification took the form of tree planting. According to the *Chinese Science Bulletin* (*Kexuetongbao*), afforestation serves as the best means to fight against the expansion of desert. Scientists were busily investigating the most resilient plants. Since 1958, afforestation in the Ala Shan Desert and the Kubuqi Desert in Inner Mongolia had accelerated.[14] Forester Cheng Chongde boasted that China, driven by the fever of the Great Leap Forward, was endeavoring to green all its desert in no more than a decade.[15] In Inner Mongolia, tens of thousands of residents were mobilized to plant trees in the desert. A local government organized five dedicated tree-planting organizations, united under the Greening Banner (Lühua qi) in eastern Inner Mongolia in 1958 to institutionalize tree planting. In a single year, they completed almost 80 percent of the tree-planting tasks scheduled for the next five years.[16] Also in 1958, the newspaper *Inner Mongolia Daily* (*Neimenggu ribao*) promoted a tree-planting campaign against the Mu Us Desert.[17] The article argues that when trees flourish, livestock thrive as well. In addition, it notes that Communist leaders were urging locals to recognize the benefits of tree planting for the prosperity of their offspring. The leaders asked residents to focus on the long-term goal of afforestation, which required peasants to incorporate tree planting into their daily lives, including negotiating the tension between farming and afforestation. Another article in the paper, "Covering Thousands of Miles of Desert with Green Cover" ("Wanlishamo pilüzhuang"), reports that the local government in Linhe, Inner Mongolia, had resolved the tension between agriculture and forestry.[18] The government also published a detailed technical handbook, providing farmers with comprehensive guidance on the selection and planting of tree species tailored to various desert environments in Inner Mongolia (see figure 3.1).[19] Focusing on the outcome rather than the process, the government organized forestry-protection crews. According to local records of the Ikezhao League of Inner Mongolia Autonomous Region, by 1979, 842 forest farms were established with trees covering 6.51 million *mu* (4,340 square kilometers) to fight against encroachment by the massive Mu Us Desert (see figure 3.2).[20] Yet another article in *Inner Mongolia Daily* claims that local cadres were taking concrete measures to ensure seedlings' survival.[21] The cadres carefully considered the possible results of tree planting, including loss of care of seedlings, inappropriate selection of tree species, and improper choice of tree-planting locations. Since 1971, the article reads, they had achieved notable success in controlling the desert by reducing the cost of tree planting and by selecting suitable tree species.

FIGURE 3.1 Cover of *Inner Mongolia Sand Control Afforestation Handbook* (*Neimenggu gusha zaolin shouce*). Source: Neimenggu linye kexue yanjiu suo, *Neimenggu*.

FIGURE 3.2 Marching into Mu Us Desert to plant trees in Shaanxi Province. Source: *China Pictorial* (*Renminhuabao*), September 1976.

Nomadic Spring: Planting Inner Mongolia's Borderland

While the *Inner Mongolia Daily* faithfully reported the government's conviction that tree planting was a positive task with universal appeal, practices at the local level were far different. Produced by Changchun Studio in 1975, the feature film *Spring in the Desert* illustrates how resistance from the ethnic population emerges when supposedly universal values are at odds with local ones. *Spring in the Desert* centers on the conflict between two Mongolian women. As a girl, Narenhua was saved from an oppressive Mongolian herd owner by the Communists;[22] proselytized by the Han Chinese–dominated Communist Party, she adopted the settled Han lifestyle. Her rival, the Indigenous leader Damulin, embraces the nomadic ethnic Mongolian way of life.

The film begins with a huge sandstorm in the Inner Mongolia grassland where horses are galloping in all directions. This sandstorm is so massive and intimidating that the film uses a long shot to capture the Indigenous nomadic life patterns against the backdrop of the boundless wilderness (see figure 3.3). Confronted with the sandstorm, Damulin attempts to migrate to escape to a more hospitable area, while Narenhua believes that staying in place and planting trees is the best solution to grapple with desertification. For Damulin, a nomadic lifestyle epitomizes Mongolian identity. By contrast, Narenhua, heavily influenced by the Han Chinese–dominated Communist Party, embodies a settled lifestyle. Empowered by Maoist environmental ideas, Narenhua regards tree planting as a universal solution that can be applied in all places, including the borderland. The local residents' environmental ideas are overlooked by the state power.

Because of her assimilation of Han Chinese environmental ideas, Narenhua's perception of the harsh natural world stands in sharp contrast to that of Damulin, who claims that the best way to avoid the sandstorm is to migrate and asks the locals to sell their livestock, move away, and return to hunting and gathering to sustain themselves. A villager who sides with Narenhua contends that "we are now members of the People's Commune: we should fight against the natural disaster collectively, and we should not run away from the famine-stricken area every time we encounter a few difficulties."[23] This villager presents a contrast between the tough individualist nomadic lifestyle before the Mao era and the collectivist and settled lifestyle in the Mao era. Natural disasters now can be defeated by the leadership of a new organization—the People's Commune. Later, Narenhua challenges Damulin by using rhetoric similar to that of the villager, claiming that "the desert could be managed well, as the mountains and water have already been managed well," if people would only stay and look after it.[24] Despite being a native Mongolian, Narenhua adopts Han Chinese ideas of the natural world, which spurs her determination to transform the desert into an oasis.

Narenhua's ideas reflect the words of Mao Zedong. In 1957, Mao delivered a famous speech in which he called for people to "unite all the ethnic groups to launch a new war—a war against nature."[25] For Mao and his followers, the Han and all the ethnic minority groups should not beg for favors from nature but should command it instead. The impact of the speech was immediate and makes its way into the film: Narenhua proudly proclaims that she has seen Chairman Mao emancipating the oppressed people from the brutal landlords and has brought back a red apple from Mao himself. The apple is circulated among the villagers (see figure 3.4). A medium shot, which depicts the smiling faces, staring eyes, and enthusiastic mood, deftly captures people's acknowledgment of such associations. More importantly, the smiling faces of Han Chinese and Mongolians in camera panning and medium close-up shots implicitly connect the ethnic Mongolians with the Han, blurring their disparate lifestyles and notions about the environment.

Convinced of their universal benefits, Narenhua brings the ideas of tree planting from the capital to the borderland. According to Narenhua, the desert represents an old-fashioned society, while the prospective oasis embodies the new Communist regime. Leaning heavily on the apple as a symbol, Narenhua contends that the best way to live up to Chairman Mao's expectations is to transform the desert into an oasis where apple trees can grow. Mongolians therefore should follow the guidance of Mao.

Claiming that during her visit to Beijing she learned of other places' successful experience of tree planting, she insists that "if we want to make contributions to the country, we must control the desert."[26] Her journey to Beijing has served as a spiritual pilgrimage for her as well as a political endorsement of her agenda. Narenhua quotes Mao's general line for socialist

FIGURE 3.3 The nomadic lifestyle. Source: Zhu and Liu, *Spring in the Desert* (*Shamo de chuntian*).

FIGURE 3.4 People admiring the apple from Chairman Mao. Source: Zhu and Liu, *Shamo de chuntian*.

construction, put forward in 1958: "Go all out, aim high, and achieve greater, faster, better, and more economical results to build socialism" (Guzuganjin, lizhengshangyou, duokuaihaosheng).[27] Narenhua's ideas epitomize Maoist political philosophy, which views nomadic pastoralism as an evolutionary dead end that stands in opposition to national progress, scientific rationalism, and economic development.[28]

As Narenhua mobilizes her people to control the desert, a close-up of a red flag reinforces the universal benefits of tree planting endorsed by the Han Chinese (see figure 3.5). The first four characters on the flag evoke a classical Chinese story of a ninety-year-old man who leads his children, grandchildren, and great-grandchildren to remove the mountains that obstruct their way in front of their houses. The old man claims that the mountains will ultimately be removed if his descendants continue the effort. The mountain-removing narrative not only embodied a settled way of life but also acquired political implications under Mao's leadership. In 1945, Mao cited this story in arguing that "today, two big mountains lie like a dead weight on the Chinese people. One is imperialism, the other is feudalism."[29] Emboldened by Mao's ideas, Narenhua calls on the local residents to regard

FIGURE 3.5 **Red flag with Han Chinese (*top*) and Mongolian (*bottom*) characters that read, "Transform China in the spirit of the foolish old man." Source: Zhu and Liu, *Shamo de chuntian*.**

tree planting to fight desertification as a sublime task. The desert thus becomes what Ban Wang calls a "sublime figure," which suggests that politics became an aesthetic process in the Mao era.[30] In the film, the desert is an object that should first be loved because it embodies a sublime task and then denounced because it is an obstacle to be removed. In this process, the Han Chinese conception of nature will come to prevail in the borderland (see figure 3.6).

The language used in the film also reflects the Han Chinese attempt to replace Indigenous perceptions of nature with their own, ignoring the localized environmental conceptions of ethnic minorities completely.[31] Eight gigantic Han Chinese characters occupy a large portion of the red flag, leaving small Mongolian characters at the bottom. The translation of the Han Chinese fable of reshaping nature into the Mongolian language entails instilling the Han perception of nature into the Mongolian mindset. In addition, Narenhua's accent-free Mandarin exemplifies the replacement of natural perceptions as she embraces Han Chinese environmental attitudes.[32] As Xiaoning Lu argues in her study of Mao-era films, the use of Mandarin "aurally pro-

FIGURE 3.6 Mongolian civilians shaping the hostile environment by going into the desert to plant trees. Source: Zhu and Liu, *Shamo de chuntian*.

motes the national unification; it also normalizes ethnic subjects into political subjects."[33] To take Lu's argument one step further, using Mandarin in this film reconstructs the environmental subjects by imposing and reshaping their environmental attitudes toward the desert. Once Mongolians speak Mandarin, they may subconsciously assimilate Han Chinese environmental notions at the same time as Mongolians begin to understand the natural world using the conceptual apparatus of the Han Chinese.

Later in the film, the promotion of tree planting encounters resistance from those natives who adhere to Mongolians' traditional pastoral lifestyles. When one villager says that "fleeing from the famine cannot contribute to socialism" and that the only way to achieve the socialist utopia is to confront the harsh environment directly, Damulin argues that "if small disaster comes, we should migrate to other pastures. If bigger disaster comes, we should move our homes."[34] For Damulin, migration to other pastures is not the same as fleeing from the famine. Instead, it represents a nomadic lifestyle that forms an essential part of his Mongolian identity. Damulin's argument is rooted in Mongolian cultural history: During the Qing Empire, the Qianlong Emperor offered Mongolians livestock and horses instead of silver and rice to preserve their nomadic identity.[35] This aspect of Mongolian identity is strongly linked to wilderness and freedom.[36] The Mongols of Northern China living beyond the Great Wall tend to regard enclosed land as a despised symbol of the cultivating orientation of Han civilization. In this sense, Damulin represents traditional Mongolians, who lead a nomadic lifestyle characterized by mobility and freedom.

The film also reveals the Indigenous people's perception of nature from a religious perspective. The Indigenous Buddhist notion of nature casts doubt on the "universal" application of tree planting by the Han Chinese. The film's depiction echoes actual local residents' environmental ideas. According to an article in *Inner Mongolia Daily* published in the Mao era, some people argued that "the desert is an incarnation of deities, and the nearby water is sacred. Thus, they should be not transformed."[37] The local Communist government responded by instructing local residents to abandon this idea. *Spring in the Desert* dramatizes this contrast through camera angles. A high-angle and extreme long shot of the masses' labor shows the enthusiastic endeavors of the people (see figure 3.7). The bird's-eye view that en-

FIGURE 3.7 **Collective efforts to plant trees in the desert. Source: Zhu and Liu, *Shamo de chuntian*.**

compasses a totality of the masses' intention directs itself toward a single purpose by the benevolent and all-seeing state. The film also employs deep focus to visualize the broad scale of the environmental crusade launched by the state power, demonstrating the widespread support of afforestation efforts from both the Han Chinese and Inner Mongolians. Nevertheless, a Buddhist called Sangbu, kneeling and praying on the ground, insists that everything in the desert is part of the Buddha's divine plan, and he feels guilty about changing it as everything surrounding him is sacred (see figure 3.8). Religious studies scholar Vesna Wallace explains how the Mongolian Buddhists have "a variety of popular folk beliefs, including the belief that unusually shaped rocks, stones, and trees or the trees belonging to shamanesses are sacred and therefore should not be touched or removed."[38] To discourage the Mongolian lifestyles and elevate Han Chinese lifestyles, colorful images of the tree-planting campaign in the daytime are contrasted by black-and-white images of Sangbu's confession at night. As a result, the film suggests that the Mongolian's Buddhist notion of the natural world is backward, with a dark future.

FIGURE 3.8 **The Buddhist Sangbu praying in darkness. Source: Zhu and Liu, *Shamo de chuntian*.**

Sangbu's religious practice epitomizes the unique Buddhist view of nature—that is, the need to preserve the natural environment. For Sangbu, the invasive Han-dominated Communist Chinese coincide with another invasive species: trees. He struggles to fight both of them at the same time. According to anthropologist Dee Mack Williams, for Mongolians, "sand functions to preserve memories of traditional Mongolian culture and thereby sustain a collective identity."[39] For the Han Chinese, by contrast, the desert means only the "primitive nature of the steppe land" and thus should be changed.[40] Ethnographer Almaz Khan offers more insight from the perspective of the language used to describe land use. For the Han Chinese, cultivation is *kaiken*, which means "opening up wasteland," but for Mongolians, land cultivation is called *gajir hagalahu*, which means "shattering the land."[41] These different ways of conceptualizing nature and land lead to serious conflict. In the film, Sangbu's Buddhist vision of the sacred land is denigrated as superstition and is further labeled "antirevolutionary," a crime that could lead to the death penalty in the name of "the dictatorship of the proletariat" in the Maoist era.[42]

At the end of the film, Han-influenced environmentalism that foregrounds the universal benefits of tree planting clashes fiercely with the native environmentalism that values local traditions. Damulin decides to end Narenhua's tree-planting efforts. He poisons the seedlings by adding toxic chemicals to the pesticides and attempts to kill Narenhua as she crosses a bridge. After surviving the attempted murder, Narenhua reconsiders the clashing conceptions of environmentalism when she encounters her fellow Mongolians. She concludes that the problem should be attributed to class struggle. In this reappraisal, Narenhua epitomizes the Communist Party's tactics in the borderland. The party consolidated its rule in ethnic areas largely by restructuring the societies in terms of class struggle and replacing their communal ties—be they religious, familial, or any other forms of ethnic bonds. Using this logic, Narenhua concludes that it is the class enemies among the Mongolians who have stopped her from reshaping the desert.[43]

The clash between the Han Chinese and Mongolian ways of life and the supposed superiority of the former are revealed in contrasting shots near the end of the film. A medium close-up, low-angle shot of Narenhua highlights her image as a powerful leader (see figure 3.9), while a long shot of Damulin makes him appear small (see figure 3.10). Narenhua's standing posture suggests a settled life pattern, while Damulin sitting on a horse visualizes a nomadic one. Narenhua condemns Damulin for wanting to return to a capitalist lifestyle by selling all the seedlings and going hunting. The film closes with Damulin suddenly being revealed as a Nationalist secret agent who was instructed to undermine tree-planting efforts in the desert.[44]

Cultural historian Tina Mai Chen suggests that Communist propaganda strengthened the power of films to elicit emotion and mass response by privileging visuality.[45] Yomi Braester further points out the practical function of Mao-era film, arguing that "the cinema was part of a larger experiment by the socialist state, aimed at producing new spatial practices."[46] The screening of *Spring in the Desert* reinforced the dominant environmental ideas and practices largely popularized by the Han Chinese. In the prominent journal *People's Film* (*Renmin dianying*), a Han student and a Mongolian student conclude that the clash between Narenhua and Damulin was intended to convey the choice between socialism and capitalism.[47] They considered Damulin, who sold livestock and actively hindered

FIGURE 3.9 AND 3.10 Pro-Communist Narenhua, shown in low-angle, medium close-up (*above top*), clashes with anti-Communist Damulin, shown on horseback in a long shot as the middle figure in the front row (*above bottom*). Source: Zhu and Liu, *Shamo de chuntian*.

the tree-planting campaign, a fictional representation of Liu Shaoqi, who had been purged by Mao for embracing capitalism. To be sure, by emphasizing the ideological conflict between socialism and capitalism rather than two distinctive modes of living, the film implicitly establishes and strengthens the bond between ethnic minorities and Han Chinese at the cost of overlooking their importantly different conceptions of the natural world. Planting trees in the ethnic borderlands is portrayed as a universally valid, wholly positive enterprise without taking the Inner Mongolians' traditions into account.

Promoted by Mao Zedong and environmental scientists, afforestation, alongside other modern conservation ideas, functions as a panacea that can remedy desertification everywhere, including China's borderland. The pretention to universality of environmental ideas is exactly what makes them inherently hegemonic. When this universalistic conservationist thought entered the ethnic borderland in Inner Mongolia, it encountered resistance, as shown in the film. Negotiating the tension between a universal discourse and a constantly changing and locally defined rhetoric brings another dimension of modern conservationist ideas—resilience—to the fore. To gain purchase in the Mao era, environmental ideas had to be recast as an integral part of revolutionary rhetoric. In *Spring in the Desert,* proselytized by the Han-dominated Communist Party, the Mongolian Narenhua represents Han environmental ideas. The conflicts between Han and ethnic Mongolians are manifested within the Mongolians themselves as a result. Narenhua strives to persuade her fellow Mongolians that planting trees instead of migrating to other places is the only and ultimate solution to desertification. The call for a uniform solution to a wide range of natural problems, including desertification and famine, eliminates conceptions of nature held by other ethnic groups in China, making the afforestation campaign inherently oppressive. After the trees have been planted, land ownership changes hands from Indigenous Inner Mongolians to the Han-dominated nation-state. Conservationist notions, when circulated, thus serve as camouflage for Han cultural assimilation and imperial expansion.

Missing Narrative: The Maoist Green March toward Xinjiang

While *Spring in the Desert* vividly demonstrates the Han-dominated Communists' ambition to green the borderland in Inner Mongolia in the form of fierce conflict, the 1965 documentary *The Army's Reclamation and Battle Song,* shot by August First Film Studio, conceals the Communists' enthusiasm to control and consolidate the borderlands behind a seemingly objective narrative.[48] While front and center in the feature film *Spring in the Desert,* the clash of viewpoints is less striking in *The Army's Reclamation and Battle Song* because of the lack of face-to-face conflict in the documentary. More importantly, while planting trees is part of the process of inculcating environmental ideas in an ethnic group in *Spring in the Desert, The Army's Reclamation and Battle Song* aims to show that the success of tree planting justifies the state's control in ethnic areas under the veneer of environmental conservation. This progress- and result-driven conservationist program is achieved at the cost of Indigenous voices.

The Army's Reclamation and Battle Song reflects the long history of Han Chinese migration to Xinjiang from the period of the Qing Empire to the Communist Revolution. In the eighteenth century, Qing China defeated the Mongol Dzungars, and many Han Chinese moved into the pacified areas. Xinjiang was incorporated into the administrative apparatus of China proper.[49] During the Republican era, Xinjiang became a political arena first for the Russian Empire and, later, the Soviet Union, the provincial government, and the Republic of China.[50] After the founding of Communist China in 1949, the People's Liberation Army took over Xinjiang and intensified its control. In 1954, under the instruction of Mao Zedong, the Xinjiang Production and Construction Corps (Xinjiang shengchan jianshe bingtuan; hereafter, XPCC) was founded to "consolidate border defense, accelerate Xinjiang's development, and reduce the economic burden on the local government and local people of all ethnic groups."[51] In 1954, General Wang Zhen, who had reclaimed the barren land in Nanniwan in Shaanxi Province in the early 1940s, was ordered to reclaim the wilderness in Xinjiang.[52] Millions of Han Chinese settled in the province: between 1949 and 2008, their percentage rose from 6.7 percent (220,000) to 40 percent (8.4 million), with the peak of the migration occurring from 1960 to 1964.[53]

Tao Zhiyue, the first military commander of the XPCC, waged a tree-planting campaign to fight against desertification in Xinjiang in 1963. Before 1966, trees newly planted by the XPCC covered around twenty-five square kilometers. *People's Liberation Army Daily* (*Jiefangjunbao*), the mouthpiece of the People's Liberation Army, claimed that the forests planted by young soldiers and young civilians from the hinterland had become a green Great Wall that locked the threatening sand in the Taklimakan Desert.[54] In addition to the XPCC, scientists from the hinterland attempted to grapple with the desert in Xinjiang. In January 1974, the *China Pictorial* (*Renmin huabao*) documented that scientists from the Chinese Academy of Sciences had arrived in the Taklimakan Desert, which was referred to as "a horrible dead sea."[55] A photograph shows the scientists marching into the desert with one holding a red flag against an endless yellow dune (see figure 3.11). The red flag indicates the central government's determination to carry out its environmental policy in the desert. These scientific expeditions sent by the Communists further facilitated tree planting in the desert by the XPCC.

According to the documentary, the XPCC served a double mission as both civilians and soldiers. The narrator cites Chairman Mao's speech

FIGURE 3.11 **Scientists marching into the Taklimakan Desert with a red flag against endless yellow dunes. Source: *China Pictorial* (*Renminhuabao*), January 1974.**

saying that "these people [XPCC] were disciplined fighters" and would become "skillful construction and emergency response units."[56] The narrator goes on to say that as soldiers, they have conquered the Nationalist enemy in Xinjiang; as civilians, they should defeat a new enemy—the harsh physical environment—by planting trees.

Local newspapers at the time reported on how the XPCC was endeavoring to reshape Xinjiang's environment by tree planting.[57] According to the *Xinjiang Daily* (*Xinjiang ribao*), soldiers were being mobilized to plant trees along the road in Shihezi. Four soldiers were dispatched to manage a line alongside the road so that the trees would look more orderly. In Turpan, children planted 49,200 trees in early March with a survival rate of 84 percent. In Kuqa, almost two thousand men and women planted a "youth friendly forest."[58] The Xinjiang government called on youth from all ethnic groups to complete the task of planting ten million trees.[59] In 1959, the tree-planting initiative expanded into the desert. In Ürümqi, the capital of Xinjiang, all the northwestern provinces and Inner Mongolia held a conference aimed at fighting against desertification.[60] One article argued that local residents adopted machines to plant trees on a massive scale, indicating the strong desire to advance tree planting.[61]

Reflecting this climate of enthusiasm for afforestation in Xinjiang, *The Army's Reclamation and Battle Song* shows that trees function as signs of the XPCC's triumph in the borderland. The beginning of the documentary portrays a serene nature that needs to be exploited by human beings. A voice-over describes the charming environment over a montage of wilderness, attractive green pastures, and alluring mountains. These beautiful natural landscapes, according to the narrator, have been awoken by the XPCC: "The wilderness has been asleep for billions of years."[62] The landscape is now shaped by human beings' sophisticated design, with straight roads and trimmed trees (see figure 3.12). Liu Mogao, a member of the documentary team, states that they purposely selected high and straight trees and used a long shot to portray the XPCC's strenuous efforts in greening the borderland.[63] The vast, wild land, according to the narrator, cannot wait to be developed further. The end of the film also eulogizes the accomplishments of Mao's tree-planting efforts through a bird's-eye view of the greening crusade (see figure 3.13) and a song extolling Mao's leadership.

The documentary's evocation of olfactory and tactile devices to depict the hostile natural environment is used to justify the needs of the XPCC's participation in Xinjiang. A close-up shot shows that the white and crystalline land is neither frosted nor snowbound but is instead covered in harmful saline-alkali soil. The acrid area is deemed the opponent of the Communist soldiers. Later, the narrator reads a poem written in the eighth century by Tang dynasty poet Cen Shen to show the historical background of the harsh natural environment:

> Don't you see how the Running Horse River flows along the
> edge of the Sea of Snow,
> Where vast and wild the brown of level sands reaches to the
> sky?
>
> The wind howls at night in the ninth month over Lun-t'ai,
> And a river full of broken boulders big as bushel baskets
> Covers the earth with careening stones blown before the
> wind.[64]

The first stanza evokes the terrifying desert sandstorms in the sky, with the second stanza describing the life-threatening boulders. In addition to nonliving things, in the film, the presence of ferocious wolves reinforces the terror of the surroundings. Depicting the environment as harsh and inhuman makes it a new target, against which Han Chinese are encouraged to take up arms immediately.

The documentary is, however, marked by a distinct absence of local ethnic people. On the few occasions when they do appear on screen, they usually seem to follow the practices of the Han Chinese mindlessly. A similar drama unfolded in North America. In his investigation of American Indian history, environmental historian William Cronon argues that "the myth of the wilderness as 'virgin' uninhabited land had always been especially cruel."[65] Cronon suggests that Native Americans were intentionally overlooked through the circulation of the discourse of "wilderness." Although Cronon focuses on the history of Native Americans, his argument also sheds light on the study of nature and ethnic minorities in Xinjiang.

While depicting a harsh environment, *The Army's Reclamation and Battle Song* barely acknowledges the historical presence of the native people in this region. This ignorance is an intentional narrative strategy to justify Communist rule in the ethnic borderland.

Ironically, while the narrator's speech employs the emancipatory rhetoric of local people, local residents' voices are largely neglected. The narrator suggests that before the Communists' emancipation of Xinjiang, it was a complete wilderness, and therefore, the land should be liberated as well. Liberation, in this sense, concerns not only the native people who claimed to be oppressed by the Nationalist regime but also pristine nature itself. The dual "liberation" of both local residents and nature indicates the oppressive policies of the Han Chinese. In this regard, the Indigenous people become what Rob Nixon, inverting Benedict Anderson's concept of the *imagined community,* calls "unimagined communities." Nixon claims that nation-states deliberately generate a narrative that allows them to overlook their own disadvantaged populations: "If the idea of the modern nation-state is sustained by producing imagined communities, it also involves actively producing unimagined communities. I refer here not to those communities that lie beyond the national boundaries but rather to those unimagined communities internal to the space of the nation-state, communities whose vigorously unimagined condition becomes indispensable to maintaining a highly selective discourse of national development."[66]

The portrayal of an empty space and of ethnic people in the documentary plays a crucial role in establishing a selective discourse of China's national development. Without giving voice to ethnic minorities, the documentary proclaims the superiority of the Communist government over its Nationalist predecessor. The narrator claims that while the previous government was unable to tame even the Tarim River, the Communist government has brought about remarkable changes by cutting down dead trees and transforming the dunes into arable land surrounded by trees. The narrator labels the transformation as achievements of the "new society."[67] Consequently, Mao's government, as demonstrated by the greening campaign, is deemed more powerful and reliable than its predecessors.

The reception of the film reveals the essence of the Communist government's tree-planting goals and its ambitions in occupying the land.

FIGURE 3.12 High-angle shot of the militarized and manufactured landscape in Xinjiang . Source: Yang and Deng, *The Army's Reclamation and Battle Song* (*Junken zhange*).

FIGURE 3.13 Bird's-eye view of newly planted trees in Xinjiang. Source: Yang and Deng, *Junken zhange*.

One viewer citing the voice-over writes that "the trees are, like their masters, demonstrating a solemn and mighty military appearance and making a brave combat gesture against the brutal sand."[68] The use of "masters" (*zhuren*) is instructive. By planting trees, the Han Chinese further justify their control of the territory. Environmental ideas are disseminated and establish a covenant between the dominant and subordinate groups. In China's borderland, this covenant is mostly a one-way street. Planting trees not only helps instill environmental ideas in ethnic groups, as shown in *Spring in the Desert*, but also legitimizes Han governance in ethnic regions under the veneer of environmental conservation, as manifested in *The Army's Reclamation and Battle Song*.

Conclusion

This chapter reveals the limitations of global conservation rhetoric when afforestation initiatives travel to China's borderland. Trees function as a green Trojan horse, indirectly reshaping Indigenous perceptions of the natural world. Promoted by Mao and environmental scientists, afforestation in the desert becomes a *green currency*—a currency that was widely accepted and circulated nationwide. Yet the circulation was by no means without problems. When tree planting as a panacea for desertification takes place in a borderland context, it may encounter resistance, as shown in *Spring in the Desert*. In this case, tree planting becomes a medium for "planting" and imposing environmental values on the local ethnic groups. In the sweeping national greening campaign across ethnic groups, the inherent cultural conflicts between the Han Chinese and the Indigenous population are downplayed. This process suggests that seemingly universal environmentalism is at odds with ethnic groups' localized environmentalism.

Despite the apparent contradiction between the environmental beliefs of the Han Chinese and those of the minority ethnic groups, each perspective may embody elements of truth based on their unique experiences and knowledge. While both sides may assert the universality of their environmental concepts, these ideas are valid *only* within specific temporal and spatial contexts. For Han Chinese state power, emboldened by forestry and environmental science, afforestation serves as the most effective and powerful means of fighting desertification.[69] Yet for ethnic groups, religious interpretations of the desert and preference for nomadic life patterns gain traction. When these two distinctive threads of environmental ideas come into contact, friction between Han Chinese and local ethnic groups becomes inevitable.

When "universal" environmental ideas endorsed by Han Chinese circulate, they operate as a potent and invisible vehicle for political control. Promoting tree planting in a harsh environment indirectly legitimizes the ownership of the land. Under the veneer of environmental rhetoric, Han Chinese exercise their governance of the territory in the borderland. In *The Army's Reclamation and Battle Song,* Indigenous ethnic groups' voices are overlooked while the XPCC consolidates its control of the borderland by planting trees. The expansionary or colonial enterprise can be implemented more deceptively by tree planting in ecologically hostile regions. This renewed framework can be found in other situations in which dominant groups employ tree planting to justify and legalize their environmental governance and political control. For example, by promoting silviculture in India, the British Empire displaced Indigenous residents' local ecological knowledge and cut off their access to land.[70] At the beginning of the twentieth century, the Japanese Empire promoted "forestry-love thought" in Korea by advancing afforestation on the Korean Peninsula as a way of assimilating Koreans into the Japanese culture.[71] Western silviculture in West Africa promoted coffee and cocoa plantations, eventually disrupting local economies, gender relations, and ecosystems.[72] Tree planting thus became a pretext for land appropriation and displacement of the Indigenous population. In China's borderland, when residents refused to cooperate with the Communists and posed a threat to the forests, the Communist government escalated the conflict and waged a battle against the enemies hidden in its forested territory—a topic to be examined in the next chapter.

FOUR

Deweaponizing Forests and Engineering Nature

Militarization, Coming-of-Age Stories, and Science Fiction in the Mao Era

IN 1954, CHINESE PREMIER Zhou Enlai (1898–1976) alerted Indian prime minister Jawaharlal Nehru (1889–1964) that the United States had parachuted spies into Chinese forests.[1] While the Communists lauded tree planting as a panacea for desertification in northern China and deployed tree planting as a means of social control, as explored in the previous chapter, the dense canopies of trees were weaponized against state power by the Communists' enemies. In the enemies' political imagination, forests had long represented an inverted social order and fluid identity. The outlaw world in the forest, both real and imagined, deeply upset the Communists—an anxiety that intensified after the Korean War (1950–53). The clash between the Communists and their perceived enemies hidden in the forest constituted an enduring fixation in the postwar period.

This chapter unveils the fact that, alongside the scientific rhetoric examined in chapters 1 to 3, the military mindset also plays a role in promoting the protection of trees and forests. It suggests that, somewhat paradoxically, the Communists' efforts to preserve forests involve eliminating threatening human adversaries—a theme depicted in Mao-era literary works. Within the context of combat, conservation extends beyond merely shielding natural resources from overexploitation, such as deforestation, to proactively protecting trees from occupation by foes who encroach on Communist territory and disrupt the societal order. The very presence of enemies drives the Communists to enthusiastically defend forests from the dangers posed by persistent adversaries. By vanquishing these foes, the Communists "purify" the forests and preserve them from Nationalist intruders, thereby reinforcing the Communists' right to the land and enabling them to interact with nature as they deem appropriate. This chapter also serves as a bridge between earlier and later chapters by exploring how conservation rhetoric during the Mao era began to infiltrate the literary realm—a trend that becomes increasingly evident in the 1980s, which will be the focus of the next chapter.

This section examines the intricate interplay between romanticized and militaristic perceptions of nature in a range of literary works related to the military from the 1950s to the 1970s, encompassing the genres of coming-of-age stories and science fiction for children. Although, except for *Tracks in the Snowy Forest* (*Linhaixueyuan*; 1957), most of these works are not well known, they uniquely illuminate the anticipation of individual subjectivity and nationhood through their portrayal of young soldiers and children interacting with the forests amid Cold War tensions. The analysis first explores the military mentality of the Communists, specifically their strategy to eliminate hidden adversaries in the forest, as portrayed in Qu Bo's 1957 novel *Tracks in the Snowy Forest*[2] and Lin Yu's 1955 short story "Songs of the Forest" ("Senlin zhige").[3] As these texts demonstrate, the Communists' engagement with the natural world and the urge to start a new order replace a romantic understanding of nature with a combative mindset, which compels the coming of age of youths, who become experienced and seasoned soldiers. Since there are very few accounts of Chinese Communist soldiers hiding in the forest to escape their enemies, this chapter takes a brief excursion into Sinophone Malaysian literature to reveal the Communists' mindset of enduring hardship and their conflicting attitudes of romanticizing and militarizing the forests. The chapter finally considers two science fiction stories that highlight conservationist concerns in a world where human enemies vanish into obscurity and nonhuman adversaries hold sway. In these sci-fi works, a more romantic and idealistic representation of nature begins to surface. This shift signifies a growing and more explicit desire to preserve the natural world. This trend becomes increasingly prominent in the literature of the 1980s.

Hide and Seek: Purifying the Militaristic Canopy

Published in 1957, the well-known novel *Tracks in the Snowy Forest* chronicles the Communists' drive to eliminate their enemies in northeastern China during the Civil War in the 1940s.[4] Zuo Shandiao, "the Eagle," the leader of the retreating Nationalist bandits, conceals himself in high mountains covered by forests. Communist scout Yang Zirong enters the forest, pretending to be a Nationalist ally. He convinces the Eagle of his loyalty and kills a whistleblower who tries to reveal his true identity. Later, when the Eagle is celebrating his success with a large party, the Communist soldiers arrive and defeat the Eagle and his men. While the conflict between the Eagle and Yang is probably the most famous part of the novel, the forest landscape functions as a narrative engine in its own right. Instead of focusing on the human conflict, this discussion will center on several less studied sections of the novel concerning the role of the forests, which empower the protagonists either to hide from or to find and engage with their enemies.

The embattled Communists regarded forests as their allies in defending themselves against the Japanese during the Second World War. As discussed in chapter 1, in the 1930s, the forester Jiang Xin pointed out that trees enabled Communist soldiers to hide from Japanese airplanes and avoid poisonous gas.[5] When the Communists were under siege, trees thus became their best means of survival. Yet the Communists' conception of forests as allies reversed itself after they assumed power nationwide. The Nationalist "bandits" in the forests haunted the Communist imagination, causing the woodlands to become places of uncertainty and anxiety and challenging the Communists' sense of ownership of the land. Some of the Nationalist forces defended their places of retreat by setting forest fires. In 1955, during the second session of the First National People's Congress, minister of forestry Liang Xi, discussed in chapter 2, articulated the gravity of the losses caused by these fires. The arson resulted in massive swaths of burned land, impairing timber production and hindering economic development.[6]

To affirm their ownership of the forests, the Communists set up the first forestry military force as early as 1948 in northeastern China. Around one thousand soldiers were enlisted to eradicate the Nationalists.[7] The Communist soldiers' march into the forest can be placed in a global context. As J. R. McNeill notes, the presence of densely wooded areas influences armies' military strategies in a number of ways: forests serve as an obstacle to movement, function as concealment, and influence societies' conception of and preparation for war.[8] In his history of Southeast Asia, James C. Scott argues that local peasants moved into the forests to avoid taxation and military recruitment.[9] Similarly, during the Second World War, Serbian guerrillas took advantage of the cover of the forests to attack the Germans.[10] In the Vietnam War, the North Vietnamese armed forces used forest canopies as platforms from which to attack US troops.[11] In response, the various regular armies tried to cut down or burn forested areas to see their enemies and reduce the chance of being ambushed.

While the United States relied on the highly toxic herbicide Agent Orange to clear vegetation and flush out the enemy in Vietnam,[12] Chinese Communist soldiers predominately came into the forest to combat their Nationalist enemies directly. In the early 1950s, the attacks from Nationalists hiding in the dense forests became a serious threat to Communist China's political stability. The People's Armed Police Forest Forces were therefore established in most of the forest-rich regions nationwide. *Tracks in the Snowy Forest* portrays a prehistory of the founding of the forces. In the novel, the Communist forces claim to be protecting the forest and the local residents, but the protection is motivated by a desire to further exploit the natural resources. This goal could not be accomplished without eradicating the banished outlaws living in the forest.

Based on their social and political roles, local farmers, Nationalist bandits, and Communist soldiers hold different views of forests in *Tracks in the Snowy Forest*. The farmers perceive the forests as innately serene places that have been contaminated by the infiltration of bandits. The bandits hiding in the woodland pose serious threats to the farmers' daily lives. The bandits come to the village from time to time and attempt to kill farmers who support the Communists. For the bandits, on the other hand, the forest is no idyll but a cursed quagmire. The bandits' perceptions of the forest reflect the Communists' political campaigns to persecute the landlords. Bandit chief

Zuo Shandiao, "the Eagle," declares, "You force me to give up my land; I will not let you live in your house. You forced me to go to the forests to starve; I will send you to the dungeon to drink dirty water!"[13] The bandits' hatred stems from the Communists' land reform during the 1945–49 Civil War, when they forced the landowners to redistribute their holdings to the poor. The land reform mobilized the masses into a state of messianic fervor, which was the core of Mao's strategy and the secret of his success.[14] It dramatically reshaped the rural society's social networks and contributed to the Communist regime's seizure of power in 1949.[15] *Tracks in the Snowy Forest* is situated in the aftermath of land reform when these formerly wealthy and powerful figures have retreated to the forests and become bandits. For them, the forests are harsh and inhospitable.

The Communists, it turns out, do not hold the forest in much higher regard than the bandits do. The soldiers see it as mysterious, frustrating, and unsettling—mostly because of the bandits. Party Secretary He warns that the former landlords, secret agents, military police, and defeated Nationalist soldiers are congregating surreptitiously in the forest. He tells the Communist troops that to ensure land-reform success, these enemies must be eliminated immediately.[16] Thus, for the Communists, the woods are a dangerous world full of enemies who must be identified and rooted out.

The forest offers an ideal site in which to avoid government control and even to build a new, inverted social order, as evidenced by folklore, literature, and historical precedent worldwide. In British folklore, Robin Hood, an outlaw hiding in Sherwood Forest in Nottinghamshire, reconstructs the social order by robbing the rich and helping the poor. Robin Hood conceals himself in the forest, where "its canopy is his hood, and its foliage his robe."[17] In *A Natural Perspective,* Northrop Frye notes that in some English plays, "the action moves from a world of irrational law into a forest. . . . The forest world presents a society in contrast to the one outside it. . . . The forest society is more flexible and tolerant than its counterpart."[18] In William Shakespeare's *As You Like It* and *A Midsummer Night's Dream,* the protagonists are determined to leave the civilized world and venture into the "green world" to escape from the irrational laws of society.[19] Similarly, in imperial China, woodlands were often used as refuges by those seeking to evade the empire's control. The famous novel *The Water Margin* depicts the marsh and its wooded areas as a liminal space where those who have been wronged by the state during the Song dynasty could seek refuge and possibly redemption. During the Qing Empire's Taiping Rebellion (1851–64), insurgents leveraged the forest's dense landscape for tactical advantage, underscoring the forest's enduring role as a crucible for resistance and societal reformation. The forest's role as a sanctuary for dissent endures, echoing a timeless legacy of nature as a catalyst for societal change.

In depicting the hunt for the bandits, *Tracks in the Snowy Forest* skillfully intertwines the challenges of military operations with the breathtaking beauty of nature. As the Communist commanding officer Shao Jianbo's team of thirty-six soldiers marches deeper into the forest, the sublime landscape evokes a romantic appreciation for the splendor of their surroundings. They become acutely aware of their own insignificance, and they experience both pleasure and awe in the face of nature's grandeur: "Who can truly know how majestic and expansive Aristocrat Range is? People say, 'Aristocrat Range, Aristocrat Range, with three thousand and eight hundred peaks.'"[20] "They gazed at the forest's wonders in the afterglow of the evening sunset. Across from them, on the branch of a large tree, there was a tree hole as big as a millstone. A large black bear climbed up and eventually entered the tree hole. The small team now lives alongside wild animals every day."[21] The novel swiftly transitions from highlighting the magnificence of nature to emphasizing the challenges posed by the harsh environment, with its biting winds and rugged mountains. These same formidable conditions inspire reverence and awe in the original inhabitants of the forest, the local villagers, who pay homage to the land that shapes their lives.

Tracks in the Snowy Forest uses dark images to present the forest's potential to challenge the Communist regime. Unpredictability and threats loom. Animals constitute one danger:

> Aristocrat Range, Aristocrat Range,
> Three thousand eight hundred peaks,
> Even the lowest no man has reached,
> And the highest has ne'er known the song of birds.

> So ran an old folk rhyme, and it wasn't much of an exaggeration. The mountains seemed heaped one upon the other. They rose through the clouds, and their tree-tops pierced the heavens. They were alive with roaring tigers and growling bears; wild boars roamed in packs. There were panther and

> deer and flocks of antelope. In the forests, you couldn't see the sky; from the heights, you couldn't see the foothills. Somewhere in that ocean of trees was a small gang of murdering bandits. But how could you find them? Where should you begin to look?[22]

The dangerous megafauna may appear suddenly and kill the young soldiers, who must remain alert at almost every second. This constant vigilance leaves the soldiers in a state of exhaustion. In addition, the forest canopy covers the entire sky, blotting out the sun in the chill of winter, and intensifies their gloomy perceptions of their surroundings. The darkness, together with the unseen lethal threat from the animals, creates an atmosphere of dread and despair.

The bandits, who reside higher up on the mountain, prove to be elusive targets and every bit as threatening as the invisibly lurking animal predators. "Seeing the thick and dark forests, Scout Yang Zirong and Sun Dade feel somewhat frightened, and ask themselves 'are there bandits around here?' "[23] Liu Xuecang exclaims, "Damn! It is hard to see anything in this forest! I can see nothing in such a short distance!"[24] The chilly weather freezes the soldiers' bodies, reducing their combat readiness and lowering their morale. After several days of searching, they find no sign of the enemy. The omniscient narrator articulates the difficulty of the hunt: "In this dense forest and in this dense bare grass, finding the enemy is like looking for a needle in the wilderness. I have searched all the hills and all the little ditches around me for a few days, but there was nothing to be learned about it."[25] Eventually, the forest furnishes some clues: broken tree branches and footprints hint that the bandits might be nearby. But the mystery-shrouded woods do not give up their fugitives easily.

The narrator says, "On the top of the mountain is an old forest. There are hundreds-of years-old, crooked woods. Some trees are towering. The ground is full of carpet-like grass, which is softly spread on the ground."[26] The towering trees are sublime, overwhelming figures, which imply the weakness and insignificance of the soldiers. The thick grass reinforces the wildness of the forest because there is no path through it. A snowstorm blows in: "The depth of the snow reached several feet or even dozens of feet, and once one falls, it is hard to crawl out. Such places are deep valleys with narrow roads. The wind blows heavy snow and fills the gully. It is not surprising to find that if one goes to such a place, the individual freezes, starves to death, or is crushed to death by snow."[27] The sensory, tactile, and visual aspects present the enormous challenges that the Communist soldiers have to suffer. If human foes are invisible, nature serves as an enduring enemy that forces the soldiers to remain constantly alert as they climb upward.

The bandits' occupation of higher ground mirrors their social position as the dominant military force in the human world at that moment. On top of the mountain, the bandits' barracks and military fort are situated inside the virgin forest; the bandits station their fighting units with solid construction and well-designed preparations against possible attack. Each company has a bunker shack with sufficient food and ammunition for combat. The layout of the military fort shows not only the sophistication of the bandits' military preparations but also the exceptional courage demanded of the Communist soldiers as they move to hit the target.

Because the bandits' woodland fortress presents such a formidable challenge, the Communists attempt to seduce the bandits to come out into the open plain, where they can kill them—especially in places where the Communists outnumber the Nationalists. They finally arrive at an open grassland without trees. The novel contrasts the plain with the forests, highlighting the military function of the trees. The narrator depicts the process used to hunt the enemies:

> For two successive days, the detachment played hide-and-seek with them.
>
> The morning of the third day, the PLA [People's Liberation Army] men climbed a high mountain. They see a large hollow bare of trees and filled with nothing but white snow. Near the northern end was a bare hill, like a huge snowman sitting in the hollow. Strange. In the vast trackless forest, where did this treeless hollow come from? Everyone examined the phenomenon curiously. After careful inspection, they decided that two or three years before, a forest fire had started in the hollow, and a west wind had pushed the flames up the hill to the east. When the trees in the hollow and on the hill had been burned clean, a sudden rainstorm had extinguished the flames.
>
> Viewing the scene, Shao Jianbo nodded to Yang Zirong. "A good place for a battle. Let's have it here."[28]

The burnt forest does not evoke a sense of darkness and threat. Instead, it is an open space with no place for enemies to hide.

The bandits attempt to capture five Communist soldiers who are crossing the hollow to hide themselves in the forest on the other side. One bandit says, "There are no forests here. The horses can run faster as a result! We will certainly arrest them! Capture them alive! If we cannot catch them alive, we kill them by shooting!"[29] Both the Communists and the bandits see the forest as an asset when looking for a hiding place and as a burden when attempting to track their enemies. The battle is intense, and the Communists eventually force the bandits to retreat back into the forest. The soldiers are now able to relax and enjoy the beauty of nature. The forest is then portrayed with bright images. Sunlight and white snow, together with the leisure of the soldiers, suggest that the forest has become not only a safe site but also a land of hope and leisure, promising a rosy future. The portrayal of romantic scenarios paints a picture of peace, harmony, and optimism, fostering a deep emotional bond with the forest. This perspective resonates with the widely held values of nature's beauty as well as the patriotic feelings of the country.

The chilly snow covers the forest and reflects the sunlight, making the forest an aesthetic pleasure in winter. "The shreds of snow and dust left in the sky have fallen completely under the sun and the land's attraction!" the narrator exclaims. "The sun's light and heat shine directly into the snow, reflecting a dazzling white glow, and the forest presents the beauty of the white world."[30] The white snow and the sunlight offer bright images of nature counteracting the gloom and low visibility that frustrates them in other scenes, motivating soldiers to continue to fight in spite of hardship.

Surprisingly, despite its focus on combat, the novel introduces some scientific concepts, particularly the ecological roles of trees and their interactions with animals.[31] The Communist commanding officer, Shao Jianbo, explains to his soldiers how brush, grass, and trees can absorb heat, thereby stabilizing temperature fluctuations. He notes that at night, wind blows into the forest owing to its lower air pressure compared to sandy regions. He also mentions that strong winds in spring prevent trees from growing, which in turn hinders birds from building nests in them. The novel highlights that in other areas, forests can act as windbreaks, but in this particular setting, the scarcity of trees presents significant challenges for the people. The interplay between the wind, trees, and other living entities suggests an emerging ecological concept that will be further explored in the following chapter of this book.

Although the local residents contend with a tough environment, the novel underscores their reliance on natural resources for their livelihood. The villagers endure the harsh conditions because the area is rich in gold and timber,[32] but the presence of the Nationalist bandits keeps them from exploiting these resources. The Communists' triumph over the bandits results in the forest becoming Communist property. The Communist military force then turns to safeguarding the workers who mine gold or engage in timber production.

Indeed, the allure of natural resources is an underlying motivation behind the battle against the bandits in *Tracks in the Snowy Forest*. In the early 1940s, the Japanese occupiers cooperated with the villagers to exploit the forest for timber. After the Japanese surrendered in 1945, the Nationalist bandits attempted to persuade the villagers to work with them, but the villagers refused, according to the novel. Angered by the villagers' intransigence, the bandits persecuted them. When the Communist forces arrived in the village, the commander, Shao Jianbo, declared that it was his responsibility to protect the residents. After making this commitment, the soldiers helped the village workers move timber out of the mountain and load it onto a train.[33] The final removal of the bandits is followed by the continued extraction of timber under the control of the Communists. It reveals that protecting the forests (as signifiers of land right) paves the ways for exploiting the forests (as natural resources).

Tracks in the Snowy Forest thus reveals the military motivation—in addition to the scientific dynamics discussed in chapters 1, 2, and 5 of the present book—to protect forests from the threats of enemies living inside them. It dramatizes the process of "deweaponizing" the forests from which the enemies may launch their attacks, pushing the romantic elements to the backdrop. The Communists "purify" the forests by killing the Nationalist bandits. The forest provides an alternative social order to the legitimate official one. To reclaim the forest is to homogenize the official order and eliminate the otherness of the forest. After the enemies have no place to hide, the forest is instrumentalized to serve the socialist purpose, eventually losing its natural wildness (which will be explored in the next chapter of the present book in Ah Cheng's novella *The King of Trees*). While *Tracks in the Snowy Forest* portrays the Communists' eagerness to eliminate their Nationalist enemies, "Songs of the Forest" depicts the coming of age of a young scout, which dramatizes the dynamic process of replacing the romantic with the militaristic understanding of nature.

Watch Out, Young Scout! The Enemy Is in the Forest

"Is there anything suspicious in such a wilderness forest?" a young scout asks his company commander.[34] The scout has grown up thinking of the forest as a pleasant and inviting place but is slowly led out of that mindset over the course of his military training, chronicled in "Songs of the Forest." Published by New Art Press (Xin wenyi chubanshe) and sponsored by the People's Liberation Army in 1955, Lin Yu's short story is a bildungsroman about a young soldier's growth.[35] In their studies of Mao-era visual arts, Xiaobing Tang and Zhu Yu demonstrate that the socialist era fostered a new vision of the landscape that helped to forge a national identity and stimulate a collective aspiration.[36] While less famous than *Tracks in the Snowy Forest*, "Songs of the Forest" provides a valuable glimpse into the Communists' conceptualization of the natural world and the process of instilling these views in youth, as seen through the personal growth of an unnamed young soldier.

"Songs of the Forest" begins with a romantic understanding of nature as shown by the appeal of birds' songs to the young Communist scout, who has been deployed to the forest. The forest boasts many species of singing birds, including the horn cuckoo, the spring-plowing-announcer cuckoo, the wagtail, and the woodpecker. With the birds' twittering and the rustling wind, the forest is an endless symphony. The beauty of the forest reminds the young man of his hometown in the chilly northern borderland.

The young soldier's love of forests is associated with his formative years. When he was a boy, he climbed trees and learned the calls of various types of birds. Later, he was seized by the Nationalist army and sent to the heartland. After the Battle of Pingjin in the Civil War, he eventually joined the Communist forces. Sent to the southern borderland, he wants to explore the unfamiliar landscape.

Seeing the forests once again rekindles his passion for nature, which reflects his innocence and desire for freedom. Before his deployment, he came across a picture portraying the snow in the forests in northeastern China and was thrilled, exclaiming that he "belongs to the forests."[37] His excitement becomes more intense when he learns that he will see endless forests once again: "I was born in the forests. I love it. And love it to death."[38] But he was first sent to an urban area, where he had to study every day; there, he felt bored and disillusioned. The contrast between urban life and the forest points to two different lifestyles. In the city, classroom learning stops him from directly engaging with the natural world. He has to follow instructions and orders. The process of learning disciplines him and restricts his freedom. In a rural setting, the curiosity of exploring the natural world in a free environment energizes him. Suddenly, a new military order assigns him to duty as a scout in a forested area in southern China, reigniting his passion. He murmurs contentedly, "Forests, I have seen forests again!" He expresses his excitement at this "homecoming" and proclaims his love of a free, natural world.[39]

"Songs of the Forest" associates the protagonist's love of nature with nationalistic sentiments. The soldier encounters little difficulty in migrating from northern China to the South because the forests of the two regions are similar. He even comes to prefer the southern forest to the northern one of his youth. He reflects that "the greening of the forests makes me itchy, and the songs in the forests are so wonderful and moving. How wonderful it is in the frontier! . . . How rich are the treasures here!"[40] Motivated by the abundance and diversity of natural sources, he proclaims that he wants to transform China's southern frontier into a beautiful garden and invite Chairman Mao Zedong to visit. He sings a song with patriotic pride:

> I am a border sentry,
> Guarding the motherland's borders.
> I love the mountains at the border,
> Forests, oh! My mother!
> I know every whisper of the wind, every stir of the grass,
> Day and night, my eyes are wide open,
> I am loyal to the duties of a sentinel,
> Watching for enemies beyond the border,
> I swear to be loyal to my country, for the country is my life!

Who can have such honor as I?
I am a sentinel of the border,
Who can have such pride as I?
I am defending the peace and well-being of the country.[41]

His company commander, however, transforms his nascent nationalistic sentiments into a more antagonistic vision of the forests. A newspaper reports that the defeated Nationalist forces in Burma have encountered a food crisis, which means that they may come into the forest area on the border to seek food. The commander tells the scout to keep his eye out for footprints. Instead of viewing the natural environment as a serene and romantic place, he tries to convince the soldier of the dangers the forest holds.

The story moves from describing the birds' songs to recounting the cheerful melodies the young scout sings. The lyrics embody the beginning of a shift from the romantic understanding of nature that opens the story to a more militaristic one: "I am the frontier scout—I protect the border of the motherland—I love the mountains in the frontier—Forests! My mother!—I can recognize the movement of the wind and grass. I open my eyes all the day—I am loyal to my responsibility as a scout—Monitor the enemies outside of the frontier."[42] The young soldier is still more relaxed than alert. He believes that the enemies are outside of the forest. His loud singing shows that he feels no need to conceal himself. His disregard of his possible exposure to the enemy by singing shows his inexperience.

The young recruit becomes more mature when the squad sergeant shows him some of the enemies' tricks. The sergeant points out that the Nationalists have been using cotton pads on the hooves of the animals that transport food to them. He tells the scout, "Young boy, are you trying to find the enemies? The enemies are cunning! The hooves of the enemies are always covered, leaving no footprints."[43] One night, other soldiers find and capture some enemy troops. This event makes the young man gloomy; he feels that he is not competent to be a scout.

The company commander encourages him to press on and urges him to transform his ideas of nature from romantic to antagonistic, as befits his identity as a soldier. The commander tells the young scout to know "how to hate the enemy and learn how to use various ways to study and monitor the enemy. It needs a lot of deliberation."[44] When another soldier sings, "Forests! My mother! I know the movement of the grass and begin to open my eyes during day and night!" the commander reprimands him: "Listen up! This is the song you often sing. How reasonable it seems! But you don't know what you are doing and don't want to think more."[45] For a young scout who sings songs, the forest is a romantic arena in which he can express his passion for nature. For the commander, the forest constitutes an analytical problem: the enemy troops in the forest are dangerous and must be found and captured or killed.

The protagonist begins to learn from his leaders. As he goes deep into the forest with his squad, his romanticization of nature starts to fade. He conceals himself in a swampy area in a heavy rain. His clothes are torn by thorny plants, his boots come off and fall down a hillside, and his feet sink deep into mud. Having failed to find the enemy, he is even more frustrated.

Still, though the company leaders urge the young scout to reconsider his surroundings in the forest, his inclination toward romanticizing nature persists. The scout is thrilled to hear the birds twittering, and he boasts that he can identify the species of the birds by their calls. He also states proudly that he can imitate many different species of birds. While his company commander reminds him to keep his eye out for anything unusual in the forest, he continues to appreciate the birdcalls. He keeps looking for anything out of the ordinary but finds nothing, which again frustrates him. His mindset suddenly begins to change, however, as the enemy looms.

The narrative mode of the short story switches to one of suspense, with dark visual images and mysterious noises. Suspecting that a rustling sound is an enemy soldier, the young scout crawls toward it and goes deeper into the forest, which becomes darker. The suspicious sound ceases. The scout thinks it must have been a bird. But the sound returns and repeats. He decides that the sound is probably artificial since it has a regular rhythm. Then, he realizes that the rhythm of the cuckoo's call is also regular. He has an epiphany: "Though he has grown up in the forest and spent half his lifetime there, he knows little of the forest, compared with his company commander."[46]

The scout becomes more proficient in combat and more experienced in reading the forest. He uses his skill in mimicking birds' twittering to deceive the enemy. Repeating a sound he hears, he receives the same sound in reply from somewhere nearby. The communication goes back and forth three times. Certain that the sound is coming from a person rather than a bird, he patiently lets the source of the sound approach, then springs from his hiding place and seizes an enemy soldier. He is now able to make use of the mimicking skills he developed as a child growing up in the forest. With this in mind, the title "Songs of the Forest" holds a dual meaning, shifting from representing the melodic bird sounds enjoyed by a young scout to the artificial bird calls he uses to lure enemies. This change reflects the scout's transition from a romanticized view of nature to a more strategic, combative understanding, marking his initiation into adulthood.

Near the end of the short story, the young soldier sings again. In contrast to his carefree singing at the beginning of the story, his songs are now imbued with revolutionary rhetoric. This juxtaposition of opposed sentiments reflects the Communist Party's manipulation of emotions to further its agenda, a tactic known as *psychological engineering*.[47] His pride in his military achievements overshadows his earlier enjoyment of nature's beauty.

Nevertheless, this romantic vision is again interrupted by military conflict. Suddenly, sounds of gunfire emerge from the forest. After the Communists quickly kill their enemies, dusk falls. The winding down of the sound of cross fire in the forest suggests the success of the Communist soldiers' military operation. The young scout can finally enjoy the beauty at ease: "Tomorrow morning, the forest will sing songs again. Tomorrow, the sound of the forest is becoming more moving and beautiful!"[48] While it seems that the militaristic understanding of nature has given way to the romantic, the recent exchange of gunfire reminds readers that the combative mindset is always behind the scenes in the forests. A deep-seated anxiety about who controls the forests resurfaces and lingers.

"Songs of the Forest" employs a coming-of-age narrative to depict the growth of a young recruit whose romantic view of nature clashes with warnings from senior mentors about the forest's dangers. Despite occasional failure to identify potential threats, constant political indoctrination leads him to mature into a Communist soldier, successfully detecting the hidden enemy through the forest's sounds. His coming-of-age experience under the guidance of his senior commanders reflects the newly founded Communist nation's efforts to assert its dominion over the natural world.

The Inverted: Malayan Communist Chinese Fighting in the Forest

Tracks in the Snowy Forest and "Songs of the Forest" showcase the Communists' heroic efforts to eliminate their enemies in mainland China. But what happens when the Communists are the ones being hunted? Few literary works or historical documents deal with this issue, but two Sinophone Malaysian works provide some insights:[49] Ng Kim Chew's acclaimed short-story collection *Memorandums of the South Seas People's Republic* (*Nanyang renmingongheguo beiwanglu*) and the lesser-known novel *Starvation* (*Ji'e*) by Jin Zhimang (1912–88) depicts Communist troops seeking refuge in the Malayan forest.[50]

The Communists in the Malay Peninsula fought an uphill battle to establish their position in the twentieth century. The Malayan Communist Party (MCP), formed in 1930 and comprised of mostly ethnic Chinese members, had close ties with the Chinese Communist Party. Influenced by the broader anticolonial and anti-imperialist movements gaining momentum across Asia, the MCP initially opposed the Japanese occupation during the Second World War. Later, it engaged in a guerrilla war against British colonial rule in what became known as the Malayan Emergency (1948–60), the focus of Jin Zhimang's novel. Although the independent Federation of Malaya emerged in 1957 and the Malaysian Communists contributed to the Japanese and British defeat, they were regarded as insurgents by both the British colonizers and, later, the Malaysian government. This forced them to continue their struggle from forest outposts. The conflict, lasting several decades, concluded in 1989 with the signing of the Hat Yai Peace Accord between the Malaysian government and the MCP.[51]

The short story "Letters from the Forest" ("Senlinli de laixin") in *Memorandums of the South Seas People's Republic* depicts Communist soldiers hiding in the forest and communicating with family members outside. The Communists are suffering from a policy of forced starvation imposed by the British.[52] In a letter to his wife, one soldier explains that the policy has affected every fighter in the forest, and some have sacrificed their lives in attempts to find food. At the end of the letter, the soldier apologizes to his wife for not being able to help care for their children.

While "Letters from the Forest" depicts suffering through epistolary exchanges, *Starvation* is a more intimate story of fourteen Communist soldiers fighting the British in the Malayan forest during the same period; at the end, only five survive. The novel is semiautobiographical: born in Jiangsu Province in China, Jin Zhimang, whose real name is Chen Shuying, migrated to Malaysia in his early twenties and fought against the Japanese and then the British. The novel is a vivid reflection of this history.

In *Starvation,* the forest is both a benign host and a cold-blooded monster that indicates hunger, unpredictability, and threats to the Communist soldiers; it is both their ally and their adversary. Their connection to the outside world and the masses they hope to mobilize has been severed by their enemies. The Communist leader tells his men, "If we do not ask the masses to support the military, our guerrilla war will not proceed smoothly."[53] Unable to contact family members, the characters in *Starvation* face an even tougher situation than those in "Letters from the Forest." They feel both emotionally isolated and materially impoverished.

As in *Tracks in the Snowy Forest,* the darkness of the forest suggests unpredictability and imminent danger: "The night in the forest is indeed dark! It is deep and thick, like a black ocean. The blackness is darker than ink."[54] Bushes and fallen tree trunks are obstacles for the soldiers navigating through the darkness. The weather frequently shifts between cloudy and sunny. With a sudden gust of wind, "the forest roars like an angry sea," becoming even more threatening.[55]

In spite of these challenges, the forest also serves as a haven for the embattled soldiers. "During the guerrilla war in the forest, the cover of the forest is like a mother's embrace."[56] It allows the soldiers to hide themselves and to heal and recover.

The sublimity of nature dwarfs the soldiers: "The forest, when viewed from the outside, is dark green with mountains and peaks overlapping and reaching the horizon. But what does it look like from the inside? . . . In addition to the deep roots, there are big, withered trees and fallen vines; besides the uprooted trees, there are netlike dead trees and growing mangosteens. . . . For thousands of years, there were only the footprints of wild beasts, with no human footprints."[57] On a cloudy day, one young soldier, Yong Xing, strains his eyes to see even a tiny gleam of light. "Yong Xing is actually going around in a mysterious forest," the narrator says.[58] He becomes hopelessly lost.

Ecocritics have coined the terms *biophilia* and *ecophobia* in discussing human beings' instincts with respect to nature. Harvard biologist Edward O. Wilson contends that there is a genetic basis for the affinity humans have for the natural world.[59] On the other hand, Simon C. Estok argues that irrational fear of nature is connected to apocalyptic and militaristic narrative modes.[60] The protagonists in *Starvation* manifest intertwined feelings of both biophilia and ecophobia.

The contrast between romantic imagination and actual practice is deepened when the characters have a more active and intimate connection to the trees. Facing starvation, some of the soldiers are so desperate that they risk their lives by climbing up a coconut or lemon tree to pick the fruit. After they arrive at the top, they have to make a difficult decision about getting back down. If they jump, they will likely be killed. If they climb down slowly, the enemy's spotlight may land upon them, and then they will also be killed. They choose the intermediate path of sliding down the trunk. The narrator says that after a soldier does so, his perception of trees changes: "When one slides down like this, the coconut tree is not the beloved coconut tree tall and thin, the treetops taking up half the sky; gently swaying, its giant fronds swaying in the wind: it is terrifying and brutally tearing his flesh."[61] Forced into bodily connection with the coconut tree, the soldier realizes the shallowness of his prior romanticized understanding: the tree now confronts him as a brutal obstacle. His clothes are torn, and his skin is lacerated. Their human enemies change the embattled soldiers' way of engaging with nature, which now makes their lives miserable.

The lack of basic necessities and the other hardships help forge the Communists' intimate ties to the land. Since they have no drinking water, they have to use their hands to dig into the soil to get water from the morning rain. The forest is cold; they try again and again to set a fire but always fail. Yong Xing, the young soldier who is lost in the forest, shivers in the rain and struggles to get warm. The silence of the forest oppresses him. No matter how much he cries for help, not even an echo comes in reply. When the other soldiers finally find him, he is dying of starvation. The soldiers dig a grave for him and tell him, "You just close your eyes and sleep in the forest of the motherland."[62] But he dies with his eyes open, and the other soldiers close them before burying him; afterward, they fire their weapons into the air in tribute. This elegiac sentiment evolves into an energetic and revolutionary impulse.[63] At the end of the novel, one soldier says, "Our comrades have shed blood on every inch of the motherland."[64] The hardships of living in the forest cultivate their attachment to the place and inspire patriotism.

"Letters from the Forest" and *Starvation* show how an abstract, romantic understanding of the forest is at odds with a militarized perception of it. It is the forest that shelters the guerrillas, and it is also the forest that threatens and starves them. While they hate the brutality of the forest, the forest also enables them to survive. While *Tracks in the Snowy Forest* and "Songs of the Forest" depict the Communists' attempt to purify the forests of their enemies in mainland China, in *Starvation*, the Malayan Communists are the ones trying to avoid being purified. While the combative vision largely overrides the romantic in these works, this change contributes to contradictory attitudes, manifesting both love and hatred of the forests, as well as hope and despair. Their reactions demonstrate that individuals' attitudes toward nature are always in flux as they adjust to changing social and political contexts. In mainland China after the Mao era, this learning-and-adjusting process in potentially dangerous surroundings occurs in works of science fiction when children venture into the forest.

Sci-fi Forest for Children: The Precursor to Modern Chinese Conservation Literature

In his study of Chinese science fiction about space exploration in the 1950s and 1960s, Nicolai Volland points to the close relationship between sci-fi and science popularization.[65] His observation highlights the impact of science fiction, particularly its roles aimed at children, in advocating for the protection of the natural world. Serving as a means of popularizing science, Mao-era science fiction that deals with nature dramatizes how the natural environment can be transformed according to the human will. Tao Bennai's 1959 short story "The Story of the Climate Company" ("Qihougongsi de gushi") depicts a firm that can eliminate meteorological disasters, such as typhoons and hailstorms.[66] In 1962, Liu Xingshi published "The Cloud of the North" ("Beifang de yun"), in which an operator in Beijing can control the weather in a desert by moving water in the air.[67] These works express a utopian imagination about the natural world but in a somewhat contentious way: human beings and nature can achieve harmony only through manipulation of the latter by the former.

Since the late 1970s, sci-fi works for young readers about forests have used scientific discourse to call upon youths to conserve the natural environment. These works deftly push the military-conflict dimension to the background while projecting a romantic and utopian portrayal of a technologically engineered nature. Incorporating conservation attitudes, they serve as a transition between the Mao-era attempt to purify the forests by eliminating enemies and the mid-1980s rise of environmental conservationist literature.[68]

Published in 1979, three years after the end of the Cultural Revolution, the story "Scout in the Jungle" ("Milin shaobing"), by Yang Xunan and Tao Jinsheng, romanticizes the power of technology to engineer nature.[69] The story begins with two boys and a girl who are curious about the forest, as well as afraid that they might encounter wild beasts when they enter it (see figure 4.1). The surprise, instead, comes from the sky. A voice from a loudspeaker on a helicopter asks them to leave the forest. The voice, coming from a robot, speaks amicably and invites them to get into the helicopter. It turns out that the helicopter and the robot are being operated by a PLA soldier via remote control. The children are delighted to board the aircraft to visit the PLA control center, where they are amazed by the technology that enables the soldiers to monitor the entire forest.

Unlike earlier literary works that dramatize threats, "Scout in the Jungle" pushes the direct military conflicts into the background. The robot plays a television program for the children, *Fighting in the Jungle* (*Milin li de zhandou*), which shows that enemies who had infiltrated from nearby countries had been found in the forest two days prior (see figure 4.2).[70] Eight bandits had been hiding among the trees. The helicopter robot urged them to surrender. When the enemies refused to give up, another robot dropped bombs on them.

Screening the conflict in the film exhibits both a continuity and a rupture with the past. On the one hand, being visualized on television rather than encountered in reality, the human enemies hidden in the forest are not a threat. On the other hand, the program shows that forests still need

FIGURE 4.1 Cover of the picture storybook (*lianhuanhua*) *Scout in the Jungle* (*Milin shaobing*), which features children who are eager to explore the forest. Source: Mao Liangying, *Milin shaobing*.

FIGURE 4.2 The children and a friendly robot watch the television show *Fighting in the Jungle* (*Milin li de zhandou*), which depicts a military conflict that happened two days previously. Source: Mao Liangying, *Milin shaobing*, 42.

to be policed by the military: the fear of the enemy lingers. While in "Songs of the Forest" soldiers monitor their territory by coming into the forest, in "Scout in the Jungle" technology does the monitoring and allows a romantic human connection to nature to develop. The forest is transformed from a mysterious territory to a decipherable and governable being and from a site of military-fueled unease to a venue for environmental instruction.

Through its focus on technology and children, which it shares with "Scout in the Jungle," the 1977 short story "Tiger Traces in the Forest" ("Milin hu zong") offers an opportunity to examine the liminal period between Mao-era and post-Mao-era environmentalism.[71] Like "Scout in the Jungle," the short story depicts a group of schoolchildren exploring an unknown forest. Upon arrival, the children are amazed to find dogs living there who can do mathematics. Suddenly, a tiger appears; when one child raises his gun to shoot the animal, he trips over a stone. His weapon slips from his hands and tumbles down the mountainside. The tiger suddenly stops before attacking the children. It turns out that scientists in a "tiger ecology laboratory" are controlling the tigers in the forest with electric collars.[72] The scientists are writing a book called *Tiger Ecology* (*Hushengtaixue*) (how the notion of *ecology* develops will be explored in the next chapter) and editing a science film entitled *Tiger Traces in the Forest* (*Milin hu zong*). They insist that the region remain a no-hunting area. To protect the animals, the scientists can press a button to make them obey.[73]

"Tiger Traces in the Forest" encapsulates the late 1970s and early 1980s spirit of environmental conservation, which envisaged a pristine natural world devoid of human enemies. This notion contrasts with the earlier politicization and militarization of forests and the elimination of adversaries therein during the 1950s and 1960s. The story shifts the forest's perceived threat from humans to nonhuman animals. While the scientists in the tale are protecting tigers in a no-hunting zone, they are doing so by subjecting these majestic creatures to their control: paradoxically, the tigers' preservation relies on stripping them of their wildness. "Tiger Traces in the Forest" uncovers a modern Chinese environmental conservation irony: the very practices and principles meant to protect nature may prove destructive when the practitioners excessively manipulate the environment to suit their whims.

Although in "Tiger Traces in the Forest" trees might seem to be merely a backdrop, their ecological roles as hosts of flora and megafauna—such as tigers—should not be taken for granted. Rather, it is a result of a lengthy historical development of ecological science in China that will be examined at the start of the following chapter. Conservationist ideas began to seep into literary creations targeting not only children but also general readers.

Conclusion

The military mindset, in conjunction with the scientific perspectives explored in earlier chapters, plays a vital role in Chinese environmental protection by shielding the environment from disruptions caused by human adversaries. Protection involves more than merely cultivating seedlings and conserving forests; it also encompasses safeguarding forest areas from encroachment by adversaries. In the 1950s fiction works *Tracks in the Snowy Forest* and "Songs of the Forest," forests are a source of anxiety that hinders the appreciation of nature's beauty owing to the threats posed by Nationalist enemies hiding in them. Once these adversaries are removed by military action, the Communists assert their power over the land and impose their vision of a purified and secure environment, leading to deforestation, a topic that will be explored in the next chapter. Later, science fiction of the late 1970s and early 1980s captures the Communists' transition from a militaristic interpretation of the natural landscape to a romantic perception that integrates with a conservation vision.

The fusion of seemingly contradictory impulses—romantic appreciation versus militaristic vigilance—breathes life into modern Chinese environmentalism. Rooted in an idealistic view of the natural world, romantic notions portray a peaceful, harmonious, and forward-looking picture. This mindset encourages a deep emotional and spiritual connection to the forest, appealing to the universal values of beauty and peace of nature among the general public. In sharp contrast, militaristic depictions present a vigilant and disciplined mindset, regarding the forest as a contested space vulnerable to exploitation and occupation by adversaries. Conservation is not merely about nurturing and preserving but also about eliminating, enforcing, and protecting. The militaristic perspective emphasizes the pragmatic urge for active defense and management of forests and natural resources, resonating with a Cold War–era mentality. The tension between the romantic and the combative as well as the free and the disciplined constitutes the narrative engine that purifies the forests examined in this chapter. While disseminating the adaptive form of conservationist notions in the context of combat, this dynamic makes the narrative oscillate between a broadly appealing aesthetic and a specific Cold War paranoia.

Understanding Chinese environmental conservation and its underlying dynamics requires a deeper probe into Chinese intellectual ideas and historical trends. During the mid-1980s, when more direct conservation rhetoric began to arise in Chinese literature, writers were taking two distinctive directions in reflecting on the roots of ecological disasters. This divergence will be examined in the following chapter.

FIVE

Great Green Awakening

Ecological Science and Chinese Conservation Literature of the 1980s

THE 1980S MARKED A period of profound change in China on all fronts as the nation began transitioning from a socialist-planned economy to a unique mix of market-oriented reforms under authoritarian governance. Against the backdrop of these political and economic changes, a historic cultural movement was gaining momentum. A mixture of idealism, opportunism, and radicalism motivated China's intellectuals and writers, who were heavily influenced by ideas from the West. Usually hailed as *culture fever,* the 1980s cultural movement was engaging and energetic: various forms of "fever"—including Jean-Paul Sartre fever, Friedrich Nietzsche fever, Qigong fever, roots-seeking-literature fever, and reportage-literature fever—gained currency as ways to reflect on China's historical trajectory as well as to contemplate its destiny after the Cultural Revolution of 1966 to 1976.[1]

In this period of cultural exploration and reimagining of the different possibilities of China's future, writers and intellectuals began to raise awareness of the drastic environmental degradation caused by rapid economic development. The 1980s saw rising levels of pollution of the Yellow and Yangtze Rivers, deforestation—especially in northern China—and increasing air pollution.[2] Among these widespread environmental harms, the massive 1987 Heilongjiang forest fire was the most devastating. Blazing for nearly a month, the fire covered ten thousand square kilometers and extended even to the Soviet Union, leaving at least sixty thousand Chinese homeless.[3] This disaster reminded the entire Chinese population of the immediate threat that ecodegradation poses to their lives and welfare. Consequently, it triggered massive concern for conserving the natural world. This environmental awareness coincided with and inspired the first modern Chinese literary movement to promote explicit conservationist notions.

Although conservation gained traction in modern Chinese literature and culture, conservation rhetoric, somewhat perplexingly, failed to rise to prominence. While the Chinese government passed the Environmental Protection Law in 1979, its eagerness to promote the economy often canceled out its conservation efforts. This chapter attempts to investigate, from a cultural rather than an economic perspective, the underlying reason why environmental rhetoric did not take hold in the 1980s. Examining two works alongside one another helps shed light on two varieties of environmentalism that characterized the decade: what the present book terms *green nativism* and *green Occidentalism*. *Green nativism* refers to the role of traditional Chinese cultural practices as a source of remedies for widespread environmental deterioration. *Green Occidentalism* suggests that the Western environmental narrative acts as a medium to present the perceived perfection of the Western world, motivating Chinese society to adopt environmental conservation practices and possibly even advocate for further political change. These two divergent and somewhat incompatible modes of conservation make it problematic to treat China's environmental discourse as a uniform whole, thereby demonstrating the fluidity and resilience of Chinese conservationist ideas. Xu Gang's literary reportage *Woodcutter, Wake Up!* (*Famuzhe, xinglai!*; 1988) and Ah Cheng's novella *The King of Trees* (*Shuwang*; 1985) each embody one of these threads of conservation thought.[4] *Woodcutter, Wake Up!* seeks inspiration from Western conservation science and popularizes it among the general public. By linking China's ecodegradation with global conservation rhetoric, it reflects the country's underlying social and political problems. In contrast, *The King of Trees* explores the possibility of using native craft knowledge to address

China's environmental problems. The novella serves as a powerful response and corrective to works examined in chapter 3 that demean and diminish the value of local knowledge and glorify the purported omniscience of the Communist Party and its leaders, as well as an aftermath to chapter 4 in which the Communists purify the forest. As it attempts to excavate China's native traditions, the novella proposes time-tested local solutions for reversing the accelerating ecological degradation. Together, these works reveal two drastically disparate forces affecting China's environmentalism in the 1980s: on the one hand, calls to integrate China into global environmental discourse were gaining traction, while on the other hand, China's own native culture offered alternative solutions to problems wrought by increasing environmental degradation. While acknowledging the possibilities of negotiations, these competing impulses targeting the same problems reveal how two different modes of engaging with nature characterize the 1980s. And this schism, in turn, largely explains why both failed to develop into major movements.

Another task of this chapter is to further investigate the rise of modern Chinese conservationist literature from scientific and global perspectives in the twentieth century. This chapter, which deals with China's 1980s environmentalism, is at pains to demonstrate that the rise of conservation literature in this decade is by no means accidental: literary writing is deeply seated within the century-long environmental scientists' efforts to engage with Western environmental science—in particular, ecology. This chapter resumes the study of modern Chinese environmentalism explored in chapters 1 and 2 and looks at conservationist notions from a more systematic perspective. An analysis of the reciprocal interplay between environmental science and literary composition provides a comprehensive view of the underlying forces that invigorate literary works focused on the Chinese environment.

When Did Ecology as a Discipline Emerge in China? Ecological Science Percolating in *Woodcutter, Wake Up!*

Xu Gang's *Woodcutter, Wake Up!* appropriates Western conservation rhetoric in its call to address China's ecological and social problems. Born in Shanghai, Xu began his studies in Chinese literature at Peking University in 1970. He later served as an editor at *People's Daily* from 1976 to 1985. After the 1989 Tiananmen Square protests, Xu was exiled to France and returned to China in 1992. His reportage literature, which melds poetic prose with realistic depictions of nature and society, has garnered several significant awards. These include the Bing Xin Literature Award for his writing on the SARS outbreak in 2003, the Lu Xun Literature Award for his exposés on Chinese deforestation in 2018, the China Book Award for writing on the history of planet Earth in 2020, and the People's Literature Award for his works on the Yangtze River in 2022.[5] *Woodcutter, Wake Up!* in particular has had a lasting impact. Over a decade after its initial publication, it won the first Xu Chi Reportage Literature Prize in 2002 and the first China Environmental Literature Award in 2003. Hailed as China's *Silent Spring*, the work captures the national pathos of ecological degradation in the 1980s.[6] Envisioning trees as the center of the natural world, *Woodcutter, Wake Up!* popularizes modern ecological ideas and employs them to disclose China's deep-seated social problems.

Woodcutter, Wake Up! functions as a textbook that aims at popularizing Western scientific knowledge of trees for the Chinese public. The text opens by suggesting that trees are a boon to the human race because of their fundamental role in defining the natural world: "Forests [are] the green kingdom on the earth."[7] Reigning over humans in their "kingdom," trees are hailed as the key to the origins of human life as well as acting as host to all species. Insects, mammals, reptiles, and birds mingle and interact in the forests. A huge variety of other plants also call the forest home and are pollinated by the insects and other creatures that they feed and shelter.

The idea of trees as a hub of all species in *Woodcutter, Wake Up!* can be traced back to Ling Daoyang's precepts examined in chapter 1. In his 1918 *Forest Outline* (*Senlin yaolan*), Ling argues that forests form the habitats of wild animals, suggesting interaction between different natural agents. The portrayal of an interactive natural world is redolent of similar depictions in modern ecological science. While the history of ecology as an academic discipline reveals a deep-seated tendency to regard trees as the center of the natural world, in *Woodcutter, Wake Up!* it also raises questions: When did ecological ideas emerge in China? More importantly, are ecological notions equivalent to conservationist ideas in the Chinese context? To answer these two pivotal questions about modern Chinese environmental thought requires a short detour to probe the history of ecological science in modern China.[8]

The emergence of ecology as a discipline began in Europe in the early twentieth century. In 1866, the German scientist Ernst Haeckel coined the word *Ökologie* as the name for "the science of the relations of living organisms to the external world, the habitat, customs, energies, parasites, etc."[9] During the nineteenth century, evolutionary concepts characterized by Charles Darwin's notion of species' mutations and replacement became the cornerstone of modern ecology.[10] At the beginning of the twentieth century, ecology was a point of view and an integrating vision rather than a specialist discipline.[11] It was not until 1935 that the English botanist Arthur Tansley coined the term *ecosystem*, naming a concept that would prove foundational to later environmental practices. A study of the ecosystem aims not only to explore the organisms in a community but also to examine the physical factors that influence living organisms. In other words, for ecologists, plants and animals constitute a single integrated system that also incorporates nonliving factors, such as climate and soil.[12]

The new term *ecosystem* was not widely used until the 1950s, an era characterized by a mixture of high hopes and deep anxieties.[13] In this period, the development of air conditioning introduced the possibility of creating climates independent of the physical environment. The new pesticide DDT promised to solve food shortages caused by insect damage to crops. Against this backdrop, British ecologist G. Evelyn Hutchinson describes recent developments in ecological science that show how humans are playing a more dominant role in the biosphere.[14] Brothers Eugene Odum and Tom Odum reoriented ecology toward a study of interlocking systems. Their approach to ecological investigation promoted the ecosystem as its central focus, incorporating the study of nutrient cycles (e.g., the carbon-nitrogen cycle) as well as of energy flow through the system.[15]

Chinese scholars welcomed the emerging discipline of ecology and introduced it into China in the early twentieth century. Published in 1930, *Plant Ecology* (*Zhiwu shengtaixue*) by Japanese-educated Zhang Ting, professor of botany at Wuhan University, and Dong Shuangqiu, professor of botany at Sun Yat-sen University, was most likely the first Chinese ecology textbook. The authors point out that the science of ecology derives from biology. They show how ecology explores the interaction between plants and their surroundings and also discuss ecology from the perspective of natural selection.[16] *Plant Ecology* argues that nonliving elements and living organisms interact with each other: for example, small changes in climate and soil may cause large-scale changes in plant life. The authors look at the various types of soil found in China, such as rocky soil, sandy soil, and clay, examining their physical properties and describing how soil composition determines which types of plants can survive.

The next decade saw the publication of around a half dozen ecological works that framed the study of the natural environment as an academic science. Commercial Press (Shangwu yin shuguan) popularized this new area of expertise, most of which was mediated through Japanese scholars. In 1933, in his *Animal Ecology* (*Dongwu shengtaixue*), Fei Hongnian introduced the ideas of Japanese ecologist Tamiji Kawamura (1883–1964) into Chinese.[17] This text is especially significant because it defines a new concept—*environmental complex* (*huanjing de fuzati*)—that describes a mode of living in which a given set of animals, plants, and inorganic entities cannot survive independently of each other. Species are more closely connected than previously acknowledged: species sharing little hereditary material nonetheless display convergent evolution of similar traits in shared environments, while individuals of a single species that are spread over a vast geographic range continue to share the same traits. For example, tigers living in Asia or the Americas may differ in their fur covering but are fundamentally similar.[18] In 1935, Yano Munemoto's *Insect Ecology* (*Kunchong shengtaixue*) made its first appearance in China.[19] In 1936, *A Short Introduction to the Plant Community* (*Zhiwu qunluoxue xiaoyin*), based on lectures by Tokyo Imperial University biologist Nakano Harufusa (1883–1972), was published.[20] Introducing Haeckel's definition and Japanese botanist Manabu Miyoshi's appropriation of ecology, this short book points out that diverse varieties of plants exist at different elevations within a given forest and explains how forests can reduce carbon dioxide in the atmosphere and affect the growth of plants outside the forest. The pamphlet argues that the natural agents within a forest engage in a dynamic interaction: for example, if the light changes, the plants will shift accordingly. *A Short Introduction to the Plant Community* contends that understanding any one species in the natural world requires studying all the factors that interact with it. Ecology becomes a nexus that connects animals and plants and all nonliving elements, including light, water, humidity, and temperature. Published in 1937, Fei Hongnian's *Animal Ecology Outline* (*Dongwu shengtaixue gangyao*) explores the interactions of animals with the physical environment as well as with plants.[21] Huang Xiuming's 1941 *Introduction to Insect Ecology* (*Kunchong shengtaixue gailun*) explores how insects interact with heat, moisture, and plants and the competing or cooperative relationships among themselves.[22]

Influenced by the new prominence of ecological studies in the West, Chinese scientists began to highlight the dynamic interaction and global systematic connection of the natural world. These new approaches emphasize the myriad ways the Chinese natural environment is connected not only nationally but also globally. A specific plant in China belongs to a dynamic system of the entire world. Understanding these complex networks requires professional expertise. Thus, scientists need to promote ecology as an important discipline in its own right.

In 1952, the *Journal of Integrative Plant Biology* (*Zhiwu xuebao*) published an article entitled "Analysis of the Current State of the Linggu Temple Forest in Nanjing" ("Nanjing linggusi senlin xianzhuang fenxi"), arguing that plant ecology should be understood using an interdisciplinary approach drawing on geology, soil science, botany, zoology, geography, anthropology, physics, and chemistry.[23] For example, one can see that the growth of the pine tree *Pinus massoniana* (*mawei song*) in the Nanjing ecosystem is determined by water, topsoil, sunlight, and geographical features. The article recommends extending this widening of disciplinary boundaries to the training of scientists as well since the nascent subdiscipline of plant ecology will demand practitioners with expertise.

In 1956, Le Tianyu, president of Beijing Agricultural University, published a seminal article in *Forestry Science* (*Linye kexue*) that examines the roles of trees using explicit ecological concepts. "Basic Principles of Plant Ecological Interactions" ("Zhiwu shengtai xianghu zuoyong guanxi de jiben guilü") deals with the interactions of natural agents, including climate, wind, soil, water, plants, and animals.[24] Critiquing Soviet scientist A. P. Shennikov's *Plant Ecology* (*Zhiwu shengtaixue*), in which one factor is prioritized, Le stresses the comprehensive interactions of all agents.[25] He further argues that the change of a dominant agent—such as forests—can affect the patterns of the entire natural system.[26] For example, he shows that when forests diminished in the northwestern part of China, the entire territory became a pasture. Le's admonition to understand the benefits of trees through the broader ecological system coincides with the global trend of studying the natural world as a dynamic system. This notion of ecology not only links various benefits of trees in a coherent and systematic theory but also offers a strong scientific incentive to prioritize the role of trees in China's ecological management.

Mao-era foresters stressed the protective implications of ecology: they fostered a dynamic, systematic, and stable view of the environment, even though they did not use the word *ecology*. As discussed in chapter 2, minister of the Forestry Department Liang Xi promoted active interaction among various natural units, including living and nonliving parts, to generate a stable system, which even predates Qian Xuesen's popularization of cybernetics in China.[27] In this system, trees serve as an active agent to prevent soil erosion in the Yellow River Valley and effectively prevent droughts and floods in the natural system, offering a stable and dynamic prospect of the natural system at large.

The interest in ecology as an intellectually productive discipline continued in the late 1950s and the early 1960s. In 1962, Beijing hosted the first plant ecology and geobotany conference in the history of the People's Republic of China.[28] One of the key issues of the conference was the study of plant distribution under particular geographical conditions. Though focused on agricultural production, the conference demonstrated that ecology was an essential component of academic research and government policymaking.

During the Cultural Revolution, the emphasis on ecological and environmental conservation did not wane. In 1972, China attended the United Nations Conference on the Human Environment in Stockholm. At this event, China celebrated its successful environmental initiatives, such as the Greening the Motherland Campaign detailed in chapter 2.[29] Furthermore, China adopted contemporary conservation concepts, such as air pollution, the implications of chemicals for public health, and the ecology of the planet Earth. Specifically, the Chinese government recognized the dynamic interplay between air, water, and soil and their collective impact on pollution.[30] This international conference became a pivotal moment in China's environmental conservation journey when China began to embrace and even affect international conservation notions.

Green Occidentalism: When the Universal Encounters a Dynamic 1980s China

The intellectual history of the environmental sciences sketched above undergirds Xu Gang's writing on the ecological benefits of trees in *Woodcutter, Wake Up!* Ecology, as an academic discipline, reconceptualizes the natural environment, touting the benefits of a broad understanding of the dynamics of the natural world even if some of the particulars remain unknown. For example, one can know what kind of organism dominates a given area without knowing the specific identity of every component. It is the foresters of the 1920s and their successors who really promoted the conservationist dimension of ecology, even though they did not use the term. This scientific development dramatically affected Xu's mindset. In an interview, he acknowledges that environmental science was essential to his writing of *Woodcutter, Wake Up!*, noting that he extensively referred to science textbooks and papers from Peking University while composing his work.[31] In *Woodcutter, Wake Up!*, forests are presented as a universal "green kingdom" on Earth, the home of all human beings and the host of all species. In this regard, *Woodcutter, Wake Up!* follows in the footsteps of twentieth-century Chinese foresters, such as Ling Daoyang and Liang Xi, who advocated for the widespread benefits of trees as they attempted to disseminate conservationist environmental ideas.[32] More importantly, Xu transcends these foresters by bringing conservationist ideas to the general public through his literary works.

Xu Gang frames conservationist ideas as a universally applicable and advanced form of knowledge coming from the West. *Woodcutter, Wake Up!* mentions a call by the Club of Rome for increased awareness of the negative impacts of modern development.[33] For Xu, deforestation constitutes the most urgent ecological threat: "As the most complex and important part of the terrestrial ecosystem, the forest's green prosperity is a symbol of the prosperity of all life on Earth. The forest is the most important hub of material and energy exchange in nature, serving as the master dispatch room of nature's changes."[34] Beyond doubt, his depiction of the interplay between material and energy in the natural world is influenced by the popularization of ecology as a coherent and systematic theory in China, as well as by the protective functions attributed to ecology.

In addition to its vast spatial scope, Xu's exposition increases its temporal scale to encompass the *longue durée*. According to Xu, trees are a panacea for ecological and social problems because they are fundamental to life. Four hundred and twenty million years ago, plants began to blanket the earth, and they have become essential to the integrity of the planet. Trees are a nexus for all human lives. "Don't believe that the forests are just forests," Xu tells his readers, for "they are a world in which there are flora and fauna and are also the coals that are formed by the forests. Forests are a huge pillar in the green world, and the others are attached to it."[35]

Xu's celebration of the knowledge of forestry and ecology on a vast geographical and temporal scale offers a platform for critiquing China's history. According to *Woodcutter, Wake Up!*, Mao-era China was ignorant of the ecological benefits of trees. In contrast to the rising ecological awareness in the West, China in the 1950s and 1960s was undergoing a destructive and chaotic revolution during which trees were cut down to fuel iron production. Meanwhile, intellectuals who were persecuted for differing from the Communist Party line were punished by being used as cheap labor in isolated rural regions. The wave of anti-intellectualism contributed to recklessly exploiting natural resources and prevented China from discovering the latest trends in conservation efforts. Only twenty years later, in the 1980s, would the Chinese learn of the activities promoted by the Club of Rome.[36] For Xu, China's afforestation was always lagging behind. *Woodcutter, Wake Up!* reminds its readers that in China, only 12 percent of the land is covered by forest, compared with 60 percent in the rest of the world. Xu further maintains that China's environmental problems are worsening at a more rapid pace than in other nations and that China should begin to take responsibility for its impact on the rest of the world.

In addition to discussing China's role in global deforestation, Xu suggests that the Chinese central government should allow more political participation and include more voices in its environmental decision-making. In doing so, he transforms an environmental issue into a political one. Xu argues that in the United States, there is not just a single voice discussing environmental issues; rather, multiple newspapers report on government corruption and uncover deep-seated problems. *Woodcutter, Wake Up!* also praises the leadership of the United States in protecting the environment. President Theodore Roosevelt supported John Muir's call for forest protection[37] and appointed Gifford Pinchot as the chairman of the US Forest Service and endorsed his work in forest conservation.[38] Under the leadership of Roosevelt and Pinchot, the National Park Service was charged with protecting over a hundred million acres of open space.[39] Xu calls on the Chinese government to adopt a similar role in environmental stewardship as well as to bring a more democratic vision to the formulation of environmental policies.

The embrace of the West in *Woodcutter, Wake Up!* was influenced by the author's reading of Western literary works. During the 1980s, Chinese writers and thinkers were quick to accept Western practices as potential solutions to China's problems, motivated in part by the end of intellectual isolation from the Western world following the Cultural Revolution. In my interview with Xu, he cited Rachel Carson and Mikhail Prishvin as his inspirations for *Woodcutter, Wake Up!*[40] According to Xu, Carson's views of the interconnectedness of nature prompted him to consider China's natural environment from a similar perspective, and her depiction of the consequences of human-caused ecological changes alerted him to China's ongoing ecodegradation. Prishvin's writings, which offer a lyrical exploration of the Russian countryside, its people, folklore, and language and track the changes in the Russian wilderness through the seasons, compelled him to develop an appreciation for nature with a global perspective.[41]

Woodcutter, Wake Up! further urges readers to deploy this global vision to situate China's own problems. In addition to the loss of woodlands in the United States, Xu points to deforestation in India and Latin America, thereby situating China within a global trend. *Woodcutter, Wake Up!* also juxtaposes deforestation with other major events across the globe, including war, military coups, the threat of nuclear war between the superpowers, assassinations and other killings, and humans landing on the moon. In this regard, China's ecodegradation is only one component of a range of events that extend beyond the borders of any single nation. Xu Gang's work thus recalls Ursula Heise's exposition of ecological awareness against the backdrop of the interplay between the local and the global in her *Sense of Place and Sense of Planet: The Environmental Imagination of the Global.* Heise argues that "ecological awareness inheres in cultural forms that are increasingly detached from their anchorings in particular geographies."[42] Heise claims that ecocriticism should go beyond the parochial attachment to a specific place and cultivate environmental thoughts that include the entire planet. Xu Gang's work not only embodies this ambition to understand Chinese environmental issues from a global perspective but also recommends pushing the global to reflect on and criticize the local.

His book is a groundbreaking attempt if we consider the century-long history of Chinese engagement with the environment from a worldwide viewpoint. In the late Qing, attitudes toward coal and other underground natural resources became significant aspects of Chinese sovereignty and its standing in the global arena as China joined the Western-dominated world order.[43] The desire to guard its natural resources with widely accepted legal notions marked the country's eagerness to merge with that order. In the Mao era, the Chinese drive to engage with the West became masculine and heroic—even vengeful—in its effort to exploit natural resources. For example, the oilfield workers' famous words "If an oil worker roars once, the earth will shake three times" expresses a bold, even horrifying, image.[44] In this rhetoric, the earth is targeted as an object not only to be exploited but also to serve as a dynamite blast that would threaten the other side of the planet—the West. A global conception of nature thus fuels a nationalistic sentiment to exploit one's own natural resources, thereby politicizing nature for China's exceptional role in championing its economic self-reliance. With Xu's *Woodcutter, Wake Up!*, creative writing about the Chinese environment transcends the nationalist framework and national borders.[45] It is a breathtaking move—one of the very first times in modern Chinese literature that a literary work openly announces a wish to join the global environmental discourse in order to protect the planet. Xu's thoughtful depiction of the forest attempts to establish a global identity and world citizenship for China.[46]

The focus of *Woodcutter, Wake Up!* on the entire human race casts China's social and political problems into sharp relief as it uses the discourse of environmental protection as a rhetorical tool against government corruption. A 1987 forest fire in Heilongjiang Province was found to have resulted largely from abuses of power by the local government.[47] Officials were bribed to abdicate their responsibility for managing the forest; as a result, the size of the forest was dramatically reduced in a short period of time. The deforestation is juxtaposed with other gloomy, even apocalyptic, events in Heilongjiang Province: local cadres permitted contaminated meat to enter the market, a chemical plant exploded, and a bridge collapsed. Xu's argument suggests that environmental issues are also inherently cultural and social problems, underscoring the need for an integrated approach that addresses these interconnected realms to effectively tackle ecological challenges.[48]

The criticism further reveals China's underlying political problems, including limited freedom of speech. On Wuyi Mountain, for example, all trees with diameters over eighty centimeters were felled. The fine for these illegal actions, however, was a mere 200 yuan (approximately thirty US dollars). The media did not cover this issue seriously, and the law was not strictly enforced because the area was far away from the political center, Beijing. Local news reports tended to skirt the seriousness of deforestation: they emphasized conservation success stories and left the impression that deforestation was not a major issue. *Woodcutter, Wake Up!* calls for a reexamination of China's political system, including an investigation of limits to freedom of speech and oversight of the government's power. Environmental protection discourse imported from the West thus became a rhetorical weapon not only against ecodegradation but also for the reformation of Chinese political failures.[49]

Xu's work reveals a stance representative of what I term *green Occidentalism* among intellectuals and writers who employ conservation discourse from the West to criticize China's political situation. In her 1995 book on post-Mao-era Chinese cultural politics, Xiaomei Chen refers to China's 1980s antiofficial "Occidentalism," in which a constructed Western Other becomes a metaphor for liberation from ideological oppression. Chen examines a 1988 television documentary, *River Elegy* (*Heshang*), which praises the West for its openness, adventurousness, and youthfulness. According to the documentary, this admiration is attributed to Western civilization's origins in oceanic travel and trade, fostering an attitude of exploration and creativity. In contrast, China's land-based defensive civilization is portrayed as fostering a conservative and backward mindset.[50] This vision of the West and China finds its way into discourse on nature: Xu's work employs what the present book calls *green Occidentalism* to manifest the universal, scientific, and even democratic vision of environmental practices. In this way, the Chinese environmental issue transcends environmental degradation itself to encourage political participation and active citizenship.

In showing this intimate association between environmental conservation rhetoric and political aspiration, *Woodcutter, Wake Up!* exemplifies one of the central ideas of the present book: the fluidity and adaptability of environmental conservation. Environmental discourse, though invoked, is appropriated by other fields and creatively evolves into new meanings. In this instance, the appropriation of conservation rhetoric may lead to its profound permutation because Xu's environmental conservation further attacks the legitimacy of China's political governance. The appealing greenness of trees and the natural world function as an open discourse and rhetorical bank, which can be bent, transformed, and channeled into different directions.

First published in the liberal magazine *New Observer* (*Xin guancha*) in 1988, one year before the Tiananmen Square crackdown, *Woodcutter, Wake Up!* embodies Chinese intellectuals' readiness to embrace the Western knowledge system and modes of living.[51] Li Rui, a member of the Central Advisory Commission of the Communist Party of China, reflected that the dire prospect of deforestation would invite the Chinese people to pay more attention to natural science. Artist Fan Zeng argued that deforestation is related to Chinese cultural traditions and that people who cut down trees to make tombs should change their practices immediately. Ma Xiang, director of the China Writers Association, suggested that *Woodcutter, Wake*

Up! turned people's attention to psychological and societal problems.[52] *Woodcutter, Wake Up!* appeared at a time when Chinese intellectuals were seeking to recover from the chaos of the Cultural Revolution and steer authoritarian China into joining a democratic and science-driven international community.

Although the vision of political reform wound down because of the 1989 crackdown, Xu's call for more attention to conservation survived, and literary writing on the environment became more or less institutionalized in the 1990s. In February 1991, Chinese writers founded the Environmental Literature Research Association (Huanjing wenxue yanjiuhui) in Beijing. Held at the symbolic power center of the nation, the Great Hall of the People, the first conference of the society gathered a number of political celebrities, including the members of the Communist Party's Central Advisory Commission Rong Gaotang and Li Desheng.[53] The society's periodical *Green Leaves* (*Lüye*), the first magazine featuring Chinese environmental literature, premiered in January 1992. The editor's opening statement defines *environmental literature* as a genre transcending traditional landscape depictions in classical Chinese literature as well as "public-hazard literature" (*gonghai wenxue*) in other national literatures. According to the statement, environmental literature aims at cultivating environmental consciousness through a broad spectrum of genres, including essays, journalism, poems, scientific works, and novels. In the same issue, the famous author Wang Meng argues that the goal of environmental literature is the shaping of the Chinese people's mindset. He contends that "human beings who have recognized the importance of safeguarding the environment have likely contributed to a significant advancement in human civilization."[54] By situating Chinese eco-degradation in a global context, Wang follows in Xu Gang's footsteps, who ushered in the environmental movement of the most populous nation at the time into the international rhetoric of conservation.

Environmental Nativism and Seeking Ecological Roots: Folk Craft Knowledge as an Antidote

While *Woodcutter, Wake Up!* regards Western environmental practices and notions as a universal currency that can be used to resolve China's ecodegradation, Ah Cheng's novella *The King of Trees* questions how these ideas can take hold at the local level and considers deploying China's native traditions and folk culture—in particular, the use of the ax—to combat its environmental problems. *The King of Trees* draws from Ah Cheng's personal encounters with the deforestation of historic rainforests in Xishuangbanna—an experience he had when he was sent to the countryside along with two hundred thousand other youngsters during the Cultural Revolution.[55] Local craft knowledge constitutes the essence of what the present book calls *environmental nativism*, which looks at China's ecological calamity from the perspective of local knowledge and experience.

Environmental nativism is associated with 1980s "roots-seeking literature" (*xungen wenxue*), which stresses local tradition and native culture as a way of understanding and steering Chinese social dynamics. Dispatched to the countryside during the Cultural Revolution, a number of writers acquired a taste for folk knowledge and Indigenous ways of life.[56] Later, responding to the cultural changes of the 1980s, they sought to renovate Chinese literature, which had been cut off from Chinese traditions. Composed in 1985, Han Shaogong's influential essay "The Roots of Literature" ("Wenxue de gen") contends that "literature has roots. And these roots are deeply ingrained in the cultural soil of national tradition."[57] Tracing the ancient Chu kingdom during the Spring and Autumn Period (770–476 BCE), he claims that the Chu culture's mystical and profound features contrast with the Confucian orthodoxy of the Central Plains.[58] Han's intention is not simply to revive local culture: at the heart of his endeavor is the belief that Chinese literature should engage in constructive dialogue with the West until it creates its own unique tradition. When translations of Western theoretical works and popular culture gained traction in the 1980s, Han and a group of Chinese writers—including Mo Yan, Zhang Chengzhi, and Wang Zengqi—realized that excavating local traditions and reassessing native culture would be essential to making Chinese literature engage with other national literatures.[59] Roots-seeking literature distinguishes itself from the more explicitly politically oriented "scar literature" (*shanghen wenxue*) and "reform literature" (*gaige wenxue*), as well as from those "pseudomodernists" (*weixiandaipai*), who were perceived as facile imitators of Western modernism.[60]

A number of novelists—including Ah Cheng—echo Han's efforts to delve into Chinese writers' works to explore Chinese identity and subjectivity. For example, Mo Yan unabashedly unveils the primitive impulse of life—a cultural root missing in contemporary China—in his 1988 novel *Red Sorghum* (*Honggaoliang*). Examining everyday customs in Shaanxi, Jia Pingwa argues that such local customs are the only way to revitalize Chinese literature. Ah Cheng's novella *The King of Chess* (*Qiwang*; 1984) recounts a young man's obsession with food and chess. His love of eating reflects his status as a poor commoner, while his mania for chess is enshrouded in Daoist philosophy. The story shows that Daoism remained alive during the Cultural Revolution and contends that secular customs should once again be enshrined in the Chinese cultural fabric.[61]

Like *The King of Chess*, which stresses local secular customs, *The King of Trees* shows how the native conception of the natural environment contributes to understanding Chinese cultural identity. As literary critic Chen Xiaoming argues, "Roots-seeking stands in the narrow gap between historicity and modernity."[62] Witnessing drastic environmental degradation in the countryside during the Cultural Revolution, Ah Cheng employed his writing to reflect the sources of Chinese ecodegradation and chart the trajectory of the Chinese environment. *The King of Trees* envisions the future by revisiting the past to recover a local tradition and address an endangered natural world.

Set amid the Cultural Revolution, Ah Cheng's 1985 novella captures the essence of the 1980s roots-seeking literature that sought to rediscover and reconnect with traditional heritage. After the trauma of the Cultural Revolution, during which people were subject to unprecedented political upheavals and fierce mutual animosities, creative writers and other intellectuals were eager to gloss over the specter of the period and focus their energies on anticipating a new life. The central government's official disavowal of the Cultural Revolution's political agenda and its emphasis on economic development further propelled intellectuals to reconsider China's historical trajectory. They revisited the past through theoretical clarification, fictional representation, and other forms of mediation. In fiction, scar literature, reflection literature, and roots-seeking literature are the hallmarks of this impulse. While scar literature recounts the physical and emotional sufferings of the Cultural Revolution, reflection literature focuses on the broader concerns of humanity.[63] Intending to investigate deeper problems, roots-seeking literature aims to excavate the cultural roots of China's social problems during the Cultural Revolution and beyond through an "inward turn."

The King of Trees begins with a group of urban youths who travel to an ancient forest to demonstrate their ideological commitment to socialism during the high tide of the Cultural Revolution. In the novella, these so-called educated youth evoke socialist science and ideology to justify their exploitation of the forest.

The local craft knowledge depicted in *The King of Trees* offers a different perspective from *Woodcutter, Wake Up!*, which embraces the universal power of Western ecological notions. In *The King of Trees*, urban-educated Li Li, who strongly believes in the universal power of textbook knowledge, stands in sharp contrast to Knotty Xiao's local craft knowledge. The difference between the rural residents and the educated youth is, in fact, first depicted through characters carrying textbooks. The educated youth are excited to visit a new environment, and they are greeted by the local cadre on arrival. The cadre asks Knotty, a small and sturdy man, to unload the students' luggage. Following the cadre's order, Knotty is surprised to discover that Li Li has included a number of textbooks in his luggage, including *Selected Works of Marx and Engels*, *Capital*, *Quotations from Chairman Mao*, and *Quotations from Vice-chairman Lin Biao*, but he carries Li Li's heavy books with ease, which surprises the educated youth. The physical strength of the rural residents dwarfs the textbook skills of the urban educated youth.

Knotty is happy to teach the educated youth his craft knowledge, which consists primarily of his ax-handling skills. When Li Li struggles to cut down a big tree, Knotty helps enthusiastically: "He softly pulled the thong back, retrieving the knife from among the leaves and branches. Searching around him, he raised his hand, and wherever the knife fell a branch or creeper would be severed. The tree jerked a few more times and then lay completely still."[64] The jerking of the falling tree recalls a dying animal, suggesting the connection among all organisms in the forest.

Knotty's craftmanship is contrasted with the unnamed narrator's ignorance. Knotty hones his ax to cut down a tree in the forest and asks whether the narrator knows how to sharpen a knife. When the narrator replies affirmatively, Knotty seems doubtful. He is correct: the narrator's blade breaks when he tries to cut down a tree, providing Knotty an opening to demonstrate his superior skills: "Without a word he picked up another sharpened knife, took it over to a nearby log, raised it slightly in both hands, and made a swift chop. Then, hunching his right shoulder, he pulled the knife out. He lifted it up, looked at the blade, and used his right hand to swing it back into the log. This done, he let go of the handle and called me over."[65] Knotty eloquently demonstrates his craftmanship and teaches the narrator how to cut down trees without saying a word.

The text's emphasis on the auditory and visual effects of cutting down trees highlights the danger of the activity and further dramatizes Knotty's superb craftsmanship. One night, a large tree is cut by the youths but does not fall to the ground because it is suspended at an angle by a mass of creepers. Knotty arrives on the scene:

> He bent his right knee lightly, leaned to the right and abruptly straightened up. In a flash the knife shot up in the air. It seemed to hover at the top of its stroke, and before we could see what was happening, a creeper had sprung up and then drifted down further off. At the same time, we heard a snap and the whole mountain seemed to quake. Hastily drawing back, we heard from a distance a series of cracks as one by one the creepers flew into the air. Finally, the great tree hit the ground, sprang back up instantly, touched down again, bounced twice more, then rolled over in the dark, finally coming to a rest. The whole world turned silent.[66]

As Knotty deftly grabs a knife and cuts down the trees without hesitation, he is so skillful and powerful that the entire mountain seems to move. Despite his expertise, tree felling still constitutes a threat. The tree snaps, cracks, and then bounces back and forth, generating loud noises that can be heard far away. Knotty remains calm, however, suggesting that he has predicted what will happen and feels at ease amid the sound and the movement of the trees. Indeed, his calm reaction shows that he is experienced in using his knife and is glad to use it in front of the educated youth. The youth have their textbook knowledge, but it is Knotty who really knows what should be done in practice.

Alongside the disparity in labor skills, the novella juxtaposes the difference between Knotty and the educated youths regarding their perception of the environment. When Knotty and the narrator get together to make a bonfire at night, they articulate different understandings of their natural surroundings. Knotty is showing the narrator the way to the lavatory when the latter stops to gaze at the mountains and the dense forest:

> I asked: "Is this a primeval forest?"
>
> Knotty gazed at me. "You don't want to piss?"
>
> "I just wanted to look around," I answered. "Is this forest very ancient?"
>
> He suddenly cocked an ear and listened.
>
> "Muntjac."[67]
>
> At this point, I became aware of some sharp cries in the distance and felt a bit alarmed.
>
> "Are there tigers around here?" I asked.
>
> Knotty scratched his belly.
>
> "Tigers? No. There are bears, leopards, wild boar, and wild oxen."
>
> "Snakes?"
>
> He had stopped listening to the noises and squatted down.
>
> "Snakes? Yes, lots of them. And pheasants, bamboo rats, red deer, and muskrats. Lots."
>
> "Wah . . . so many different animals! Should make good hunting," I said.
>
> He got up again, turned to look at the fire in the distant clearing, and sighed.
>
> "Soon be none left. None."
>
> "Why?" I was puzzled. . . .
>
> Knotty suddenly fixed his eyes on me in the moonlight.
>
> "Have you been sent here to clear trees?"
>
> "No," I answered after thinking it over. "We're here to be reeducated by the poor and lower-middle peasants, to build up and defend our country and to eliminate poverty and ignorance."
>
> "Then, why cut down the trees?"
>
> "We'll cut down useless trees and replace them with useful ones," I replied. (We had been given a general idea of our work when we got here.) "Is felling easy?"
>
> He lowered his head. "Trees can't run away."[68]

The narrator is ignorant of the bountiful natural resources at hand. For him, the world in front of him is unpredictable and even life-threatening, full of dangerous predators. Knotty, however, appears to be aware of the natural resources of the forests. Instead of answering the narrator's question regarding the longevity of trees, he makes one more attempt at identifying the animals and explains the richness of animal life in the forests, suggesting that the forests are valuable not only for their trees but also because they provide a perfect habitat for different kinds of wildlife. The planned felling of the trees by the educated youth distresses him.

Ironically, the youth who have been ordered by Mao to be educated in the rural regions actually try to educate the villagers, including Knotty. Instead of seeing themselves as students of the rural residents, the educated youth arrogantly regard themselves as instructors. The narrator justifies cutting down trees by contending that they are, in fact, planting trees. He first sublimates the forthcoming deforestation, arguing that this crusade epitomizes socialist class liberation and promotes national security. After realizing that the argument is not entirely convincing, he later claims that new *useful* trees will replace the *useless* ones. The looming deforestation renders Knotty gloomy. He lowers his head in dejection, mirroring the felling of the doomed trees and registering his sorrow.

Why are some trees useful and others purportedly useless? The difference, according to Li Li, lies in their value for socialist construction. Only specific types of trees are regarded as useful by the educated youth. No longer would trees be classified by biological taxonomy; instead, socialist values deputize the educated youth to evaluate the natural environment according to its instrumental value. The narrative strategically overturns Zhuangzi's ancient parable that celebrates the survival of trees deemed "useless," spared from the ax due to their perceived lack of utility.[69]

The opposition between Li Li and the local residents centers on whether to fell the biggest and oldest tree in the forest. The local residents' awe of the tree stems directly from their daily interaction with nature. When the educated youth come to the mountain, the boys are enthusiastic about slashing away at the wild plants. The local cadre, however, admonishes them not to chop down the wild plants freely. The cadre says, "Stop slashing at everything and there won't be so many insects."[70] The cadre conceives of the forests differently than the educated youth: the flora and fauna have formed a natural balance, and people should not attempt to disturb it.

Here, the depiction of the interconnection between the human and the nonhuman world is germane to the trope of invasion and its subsequent erasure of local folk knowledge. The educated youth, as intrusive newcomers from the urban to the rural areas, do not realize the intertwined relations within the nonhuman world. Growing up in big cities, they do not have the sensibilities to perceive the natural environment with appreciation. For them, the plants are merely an array of isolated objects. It is the local cadre who propels them to reconceptualize the invisible and subtle interactions inside the forests. But his success is limited; though trying to preserve the ecological balance and reorient the youths' understanding of the forests, the cadre nonetheless agrees that the local residents must clear fifteen hundred acres of farmland to plant "useful" trees.

Ironically, Li's commitment to the socialist mandate, which provides his reason for cutting down the big tree, is inherently unscientific, bordering on superstitious. To justify his dismissal of the local residents' beliefs as superstitious, Li evokes a scientism that leads him to pursue progress, power, and technological control. He employs crude evolutionary concepts when he claims that the new will supersede the old. He asserts that the big tree's "location is not scientific."[71] Although the local farmers have firsthand knowledge of nature, Li hardly takes their opinions into account. His notion of science is based on textbook information driven by a uniform and oppressive political doctrine, which stands in sharp contrast to the local, intimate, and experiential way of relating to the natural world.

Li's substituting new "useful" trees for ancient forests and defining what is and is not scientific manifest two contested versions of environmentalism: so-called science-oriented instrumental environmentalism and folk-knowledge-oriented environmentalism. Previous studies have investigated Li's utilitarian environmental ethics,[72] which are driven by faith in science and a socialist-textbook mindset. The political works of Karl Marx, Friedrich Engels, and Mao, together with his own definition of science, contribute to his belief in the universal validity of his environmental values that can override all local understandings of nature. Unlike Xu Gang's *Woodcutter, Wake Up!*, which employs universal Western environmental notions to call for political reform to make China more democratic, Li's belief in the universal application of environmental values justifies and strengthens authoritarian control—a move that is redolent of the Han Chinese endeavor to promote tree planting in ethnic borderlands discussed in chapter 3. For Knotty, the forest has a unique history, and the gigantic tree is distinctive. The forest cannot be reduced to an undifferentiated collection of trees. Instead, from Knotty's perspective, it is mysterious and worthy of respect because it is home to many species.

If *Woodcutter, Wake Up!* embraces science as a universal currency that can be applied almost everywhere (much like the tree planting in ethnic borderlands considered in chapter 3), *The King of Trees* stresses that local craft knowledge plays a vital role in understanding the natural world. The local cadre believes that the insects will flourish if people do not ax the trees randomly in the forest, implying that local residents probably understand the big tree's ecological role in safeguarding the balance of the surrounding environments. This firsthand humanistic and experiential folk knowledge is at odds with Li's knowledge of the forest derived from textbooks and socialist doctrines.

As Li successfully pushes his socialist doctrine of progress, Knotty becomes a minor figure. Drawing on his own craft knowledge, Knotty tries to persuade Li not to cut down the big tree but to no avail. The cadre tries to comfort the sorrowful Knotty by reminding him of the broader revolutionary context and the inevitability of felling the tree. The cadre claims that not even he, the local party leader, can control the youths who claim to be loyal to the party. As a result, Knotty—who had been a heroic figure on the battlefield in the 1960s but was discharged from the army because of a careless mistake in disciplining his subordinates—is now dismissed again.

Knotty's "tree-cutting craft skills" (*kanshu de shouyi*) had earned him the title the King of Trees. Originally, the narrator had assumed that the King of Trees referred to the gigantic tree that they are going to cut down. His nickname, Knotty (Geda), which suggests a tangled string, symbolizes his intertwined relationship with the tree. Knotty tries to protect the gigantic tree, which becomes an iconic symbol of the natural world.

Knotty's self-destructive attachment to the tree exemplifies the Freudian notion of melancholia, a state in which the sufferer can no longer clearly define the borders between his or her own subjectivity and the existence of a lost object.[73] Ursula Heise further suggests that melancholia over lost nonhuman beings is an integral part of the environmental worldview that manifests itself in political emotions.[74] Unable to stop the tree from being cut down, Knotty suffers from melancholia, showing his intimacy and empathy with the gigantic tree. He sits down, gazing ahead, motionless. His hands loosen their grip, and the knife he wrested from Li's hand drops onto a root of the tree. At the sound, a dozen birds who made the tree their home shoot out from its leaves like arrows; making crying noises, they disappear into the sky. This scene underscores the ecological interconnectedness of the tree with other life forms.

The synchronic changes of Knotty and the tree suggest that the former's melancholia stems precisely from the porous boundary between the big tree and himself, thereby paying tribute to Knotty's bygone ways of craft knowledge in which the natural and human are organically interwoven. Knotty stands behind the tree for four days simply to watch it being felled. Although his wife brings food to him, he eats very little. Gradually, he becomes ill: "Over the past four days, his hair, streaked with gray, seemed to have grown much longer and stood up on end. His face was covered with wrinkles, which became denser nearer his forehead and ears; his upper lip was taut, the lower one slack; the skin on his neck hung loose."[75] Here, the novella chronicles the loss of vitality of both the tree and Knotty. When the tree withers, Knotty's face and hair manifest aging, and his hair becomes increasingly unkempt and disheveled, reflecting his inner agitation. These corresponding changes between the tree and Knotty problematize the boundary between human beings and the nonhuman world, orienting the narrative arc to the visual and sensory impact of this environmental degradation at hand.

The calamity caused by cutting down the tree is depicted by a shift into magical realism that borrows religious imagery to dramatize the scale of the environmental degradation. After the tree is felled, the youth set fires in the forest to clear a space. When the fire becomes uncontrolled, a muntjac suddenly appears, only to return to the burning forest rather than come out of the blaze to meet human beings. The forest becomes an inferno. The flames cover a huge area and devour the trees almost completely. The trees are portrayed as tinged with a religious aura. Uprooted, they leave the earth and rise to the sky, evoking the supernatural beings of Daoism. Meanwhile, the mountain is no longer a mystical and serene object with its own rules and balance. Instead, it seems to have become an angry monster, emitting strange noises. Though the heat is increasing, some of the people, paradoxically, feel cold. Nature regains its sovereignty by self-destruction.

As he witnesses this great derangement of the natural world, Knotty's health worsens. He is weaker each time the narrator comes to visit him. The conflagration seems to have burned up his spirit. Eventually, he dies, with his upper lip flat, the lower one sagging and baring a few teeth, his deflated face indicating his despair. The village leader makes a coffin from thick wooden boards and buries him adjacent to the big tree's stump.

After the villagers bury Knotty, the narrator focuses on the recovery of the natural environment:

> Large quantities of ash from the mountain flowed down in the rain and accumulated in a thick black layer. All day long an acrid smell filled the gully, our eyes watering from the fumes. After the rain stopped we went up the mountain to work again. The whole slope was bare save for a number of bigger trees that had not been completely destroyed by the fire—the black erect trunks with their charred branches looked like arrows from outer space which had plunged deep into the naked body of the mountain, leaving their black tails above the ground. . . . The week's downpour had brought out tufts of grass here and there, short blades of yellowish green.[76]

The barren land with the black trunks is appalling to the villagers, who are accustomed to the green, natural world. These horrifying images start to recede as nature begins to recover from the conflagration. Then, the root of the felled giant tree embraces Knotty's coffin: "In the distance, we could see that Knotty's grave burst open: the white coffin was raised high above the grave, gleaming in the sunshine. . . . It turned out that a tangled mass of shoots was growing vigorously from the earth in which the coffin had been laid."[77] The tree has lost its own body and now attempts to incorporate Knotty's. The shoots grow extensively inside the coffin, leaving Knotty's corpse bound up with the remains of the big tree. The "human world" represented by the corpse of Knotty and the roots of the tree physically intersect. Previously, it was Knotty who protected the tree; now, the tree envelops

Knotty's body, forming an organic whole from the two corpses. In this supernatural transformation, not only does Knotty incarnate the tree—as the title *The King of Trees* suggests—but also the tree comes to embody Knotty. Sadly, this kinship is only established after their mutual demise.

The novella ends with elegiac symbolism, reiterating the intimacy between the two Kings of Trees:

> The team leader suggested cremation to Knotty's widow, and in the end she agreed. On the summit we built a pile of kindling as high as a man, placed the coffin on top, and lit a fire underneath. The flames slowly mounted, producing dark clouds of smoke when they touched the coffin. There was no wind that day, and the smoke rose straight into the sky, formed a ball, paused, and then rose straight up again until it disappeared.
>
> Knotty's remains were buried in the original grave site. As time went by, a patch of grass with white flowers grew over it. People who understood these things told us that the plant was a kind of medicinal herb especially effective in healing wounds. As we continued our work on the mountain, we'd often stop and gaze across the landscape: we could see the huge trunk, scarred like a man who had fallen, and we could also see the patch of white flowers like bones exposed in dismembered limbs.[78]

The ending is an elegy for Knotty's bygone way of understanding nature. The fallen tree and the deceased Knotty have healed each other. The white flowers, as a medicinal herb grown from Knotty's bones, may be able to heal the tree's wounds. This affinity is reflected by similes of bodily reconciliation: the tree is compared to the scar of a human; the white flowers are depicted as the white bones of human beings. Even disintegrated into ashes, Knotty embraces the tree and begins to heal its wounds.[79] Driven by his grassroots folk knowledge, which stands in sharp contrast to the seemingly universal textbook knowledge of the youth, Knotty becomes a martyr to the inevitable invasive culture of Mao's followers. It is Knotty's folk knowledge that confounds the boundary between him and the tree when they physically and metaphorically bond together in a breathtaking and heartbreaking manner.

While featuring Chinese native and folk traditions, the novella exemplifies the modern concept of the ecosystem discussed at the beginning of this chapter: trees function as a hub and haven for other species, forming an interactive system. While *Woodcutter, Wake Up!* shows more clearly the impacts of Western environmental notions, *The King of Trees* also relies on modern ecological concepts that are largely indebted to the foresters and environmental scientists who were examined in the first two chapters of this book and the beginning of this chapter. In other words, although Ah Cheng may not be aware of environmental scientists' influences on him, their notions of the ecological system are deeply embedded in the novella. The story suggests that trees, especially the gigantic tree that is the centerpiece of the tale, dominate other natural agents nearby, building a stable and interactive relationship with their surroundings. The local cadre urges the students not to randomly destroy the trees, bushes, and grass since they are intertwined. He explains that if they refrain from hacking away indiscriminately, they will not be bothered by insects. The novella also exhibits the invisible functions of trees, which are usually overlooked. When the tree is about to be cut down, the birds residing in the tree flee. When the forest is on fire, a deer who lives in the forest appears. The sudden appearance of these creatures presents visual evidence that trees and forests function as hubs for all types of natural agents. The novella also dramatizes how trees can respond to disturbance by the human world. The fire accelerates after trees have been cut down and laid on the ground. Trees no longer serve as buffers for the humans but instead fuel the fire. Yet the remaining uncut trees refuse to become passive entities to be burned and assume their agency by uprooting themselves. In doing so, they protect the natural world by retreating from a dangerous site. Reflecting the modern ecological concepts, which center on the pivotal role of trees in the natural world, explored in the first part of the chapter, *The King of Trees* illustrates how modern environmental science percolates into literary works.

Conclusion

The King of Trees and *Woodcutter, Wake Up!* illustrate two distinctive and contentious trends of environmental thought in the 1980s. While the former centers on traditional craft knowledge originating in an agrarian society, the latter urges Chinese readers to learn from the West and join in the global conservation discourse. This chapter shows that the two trends, the inward (green nativism) and outward (green Occidentalism), reflect the contestation of environmental rhetoric when it circulates and adapts to disparate social conditions. Environmental discourse in 1980s China is not a unified whole. Politicized environmental rhetoric further reveals permutations that depart from their original intentions of ecological conservation. A yearning for global environmental-protection discourse in *Woodcutter, Wake Up!* and an eagerness to revisit China's own folk knowledge of nature in *The King of Trees* call for a reevaluation of China's historical trajectory and its deep-seated social and political problems.

This chapter unveils an additional facet of contested environmentalisms, juxtaposing the environmental mindset and tumult of the Mao era with the post-Mao period of reflection and remediation. While chapter 3 exposes the overbearing nature of Chinese state authority, this chapter elucidates how both local and Western approaches could have served as countermeasures to the environmental perils and political turbulence of the Mao era. Though the manifold expressions of Chinese environmentalism are still not consolidated into a unified movement, they nonetheless provide a respite and an array of tools for the cultural reengineering of Chinese society. In this way, they cultivate a covert force in opposition to the seemingly indomitable Leviathan of Communist state power.

Furthermore, both trends suggest the strengths and limits of environmentalism in the 1980s. Green nativism and green Occidentalism not only show the heterogeneity of environmentalism during that decade but also reveal the political and social dynamics of environmental rhetoric. Unfortunately, because green nativism and green Occidentalism took vastly different directions, these two voices were not able to converge to raise citizens' awareness of environmental degradation. Without a coherent voice, environmental management either becomes a byproduct of the West or is reduced to revisiting bygone traditions. Neither course is capable of producing rhetoric acceptable to the wider public. Treating Western science as a universal currency may not resonate with the grassroots masses, who are influenced by local traditions. Meanwhile, stressing the local folk culture might not appeal to the urban intellectuals who have accepted Western science. In other words, neither thread tends to approach, negotiate with, or incorporate the other in a cohesive framework that might counter mounting environmental decline. Nevertheless, and somewhat counterintuitively, this weakness might be a strength because these multiple Chinese environmentalisms have the potential to enlist a vast spectrum of the audience to ignite social change—a point that will be elaborated in the final chapter of this book.

When Chinese environmental discourse does integrate the local and the West, it can trigger a public outcry for environmental conservation. The problem, it turns out, is where environmentalism intersects with power grabs. Chai Jing's 2015 documentary, *Under the Dome* (*Qiongding zhixia*), which introduces Western conservation practices to address Chinese local air pollution issues, gained immediate traction upon its release. Within a week, the government removed it from all websites. Through such censorship, China's authoritarian government establishes the boundaries of promoting environmental conservation: as long as it does not pose a threat to the state's vested interests, environmental discourse gains currency. Otherwise, this discourse is suppressed, just like other social movements in contemporary China. While political factors weigh heavily against conservation, economic elements add another layer of challenge to Chinese environmental ideas. In the twenty-first century, with consumer culture dominating China's market and large corporations attempting to seduce users to participate in tree planting, China's consumer environmentalism makes understanding Chinese conservation ideas increasingly contingent. This will become clear in the case of Alibaba's *Ant Forest*, to which the discussion now turns.

CODA

Arbor Day Is Every Day!

Virtual Arbor Day and Consumer Environmentalism

Introduction

In 2016, precisely a century after the inaugural observation of China's Arbor Day, China's e-commerce giant Alibaba launched its tree-planting promotional campaign through its cellphone app *Ant Forest* (*AF*). Without ever visiting a planting site, a user of *AF* can participate in tree planting simply by using the app: Alibaba will plant trees on *AF* users' behalf if they record their participation in several environmentally friendly activities. The program took off quickly between 2016 and 2019. Within three years, the number of users had increased to five hundred million, and Alibaba had planted two million trees.[1] Ma Yun (Jack Ma), then chairperson of Alibaba, pledged that his company would continue to plant one million trees per year. Thus, a century after the first Arbor Day in the Republican era, tree planting has entered a new phase in contemporary China. Every day, users of *AF* help afforest China simply by engaging in environmentally friendly activities, such as riding a bike or reducing their use of plastic bags. Like the foresters and politicians who attempted to advance the conservationist agenda in the first half of the twentieth century, Alibaba has been engineering a novel perception of the natural world through its cellphone app.

While Arbor Day had been a single day since the Republican era, it has now become a daily practice—a virtual Arbor Day.

Alibaba's *AF*, a branch of its subsidiary Ant Finance, embodies the spirit of virtual Arbor Day by activating grassroots-level participation. Alibaba explains the rationale for using Ant Finance: "We would love to travel with even more little friends. We are like ants: though they are small, they work together and radiate amazing power and never give up on the way to their destination."[2] As a division of Ant Finance, *AF* inherits this spirit. The eponymous ant conjures an image of a body of anonymous small agents who can contribute to dramatic environmental improvement through their collective efforts. While the name does accurately capture the need to instill knowledge of the environment in the general public, it also cannily masks a capitalist impulse to profit from these everyday consumers when they use businesses sponsored by *AF*'s parent companies, Ant Finance and Alibaba.

This concluding chapter explores the mechanism of *AF* through a hybrid, close reading. Representative of new media more broadly, the cellphone app *AF* is an inherently multimedia text, seamlessly integrating textual, visual, and audio elements to convince its users of the benefits of tree planting. Initially, I examine how *AF* frames itself as a computer game through an interactive reward-cycle mechanism. Then, I probe how *AF* uses affect to create social equality and promote active citizenship, imbuing its users with warm and caring sentiments. In this process, *AF* reduces the requirement for environmental expertise to promote afforestation. Hidden behind these emotional and cognitive rewards for the consumers, however, lies Alibaba's own profit-driven business, possibly intensifying instead of alleviating ecodegradation. Through this analysis, I tease out how *AF* illuminates a new form of environmentalism in contemporary China—*consumer environmentalism*—which shows how Chinese conservationist ideas adapt themselves to survive and thrive. Ultimately, I conclude this book by arguing that the contestation of Chinese environmental attitudes is the strength of Chinese authoritarian environmentalism. By revisiting the book's three key contributions, the final section illuminates how studying contested modern Chinese environmental attitudes can lead to a better understanding of the Chinese nation-state for effective policymaking.

Gaming Arbor Day and Commodifying Conservation

Integrated with Alibaba's consumer app, *AF* is becoming an essential component of the firm's consumer ecosystem. It functions as a game, seducing its users into continued participation by rewarding their online achievements. As more and more people use the app, the growing pool of members also establishes a community that fosters environmental conservation. In this process, *AF* blurs the boundaries between the virtual and the real. In encouraging tree planting nationwide, *AF* promotes ordinary individuals' conservationist ideas through this virtual Arbor Day in a way reminiscent of the original Arbor Day ceremony in the Republican era: the 1916 Arbor Day was a tree-planting ceremony; virtual Arbor Day starts with a tap of a finger on a cellphone screen, which produces scientific knowledge about and affection for the natural world.

AF is first and foremost a consumer app. It is part of Alibaba's flagship app, Alipay. By clicking on the Alipay app, users can deposit money, pay utilities, hire an Uber, or order food (see figure C.1). By clicking on the *AF* app, users are invited into a well-designed "natural" world.

AF takes the form of a cellphone game and invites users to stick with it.[3] It encourages them to finish tasks they have set themselves on the tree-planting platform and provides bonuses if they reach their goals. These tasks are associated with their daily activities, which include riding a bike, reducing the use of private cars, refraining from using plastics, and recycling trash. To encourage green travel, it prompts users to walk, use shared bikes, or take subways and buses. All these real-world activities, when logged online, contribute to a bonus calculated by grams of "forest energy" that can be used for tree planting. For example, when consumers use their own mugs instead of the paper ones provided by Starbucks, they are rewarded with thirty grams of forest energy (see figure C.2). If they do not use a straw, they receive another five grams of forest energy. According to Alibaba, the rewarded grams are equivalent to the carbon dioxide absorbed by a single tree. Once a single user accumulates a certain number of grams, Alibaba will plant a corresponding real tree in the deserts of China. Thus, users reduce their carbon footprint by engaging in environmentally friendly activities, and carbon dioxide is further reduced via Alibaba's planting of a tree on

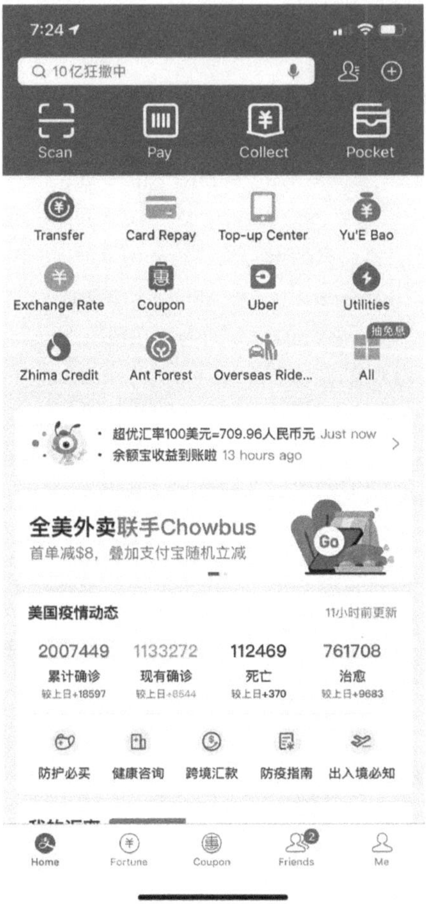

FIGURE C.1 **English version of *Ant Forest* in Alipay. *Ant Forest* is listed with other consumer functions of the app. Source: Alibaba, *Ant Forest*.**

their behalf. Throughout the process, users are encouraged by in-app messages to continue to accumulate grams. Once they receive a few grams, they are hungry for more. As a result, *AF* creates a reward cycle that aims to encourage its users to stick to the app even to the point of mild addiction. In exploring the social implications of popular culture in Nazi Germany and the American film industry, Max Horkheimer and Theodor W. Adorno suggest that the culture industry keeps consumers from being aware of social problems. When consumers put things out of their minds and forget suffering, they become powerless.[4] *AF* contravenes this understanding of the culture industry by merging entertainment with environmental initiatives.

FIGURE C.2 **The *Ant Forest* interface is designed as a computer game. Users can accumulate "grams" through daily activities to finish the task of planting a real tree. Source: Alibaba, *Ant Forest*.**

AF not only plants trees but also cultivates users by employing a feedback cycle through visual materials. Similar to many games in which users and computers interact, *AF* also animates the growth of a tree when a user conducts environmentally friendly business with Alibaba. Seeing these images of growing trees contributes to users' pleasure of control and feelings of mastery.[5] In this virtual app, developers of *AF* transform users' pleasure into growth. Like the coming-of-age novel discussed in chapter 4 in which the platoon leader instills his conceptions of the natural environment in his adolescent trainee, *AF* also seeks to imbue its users with environmental ideas.

After users have accomplished their tasks and collected grams, they may select their own tree-planting locations and species, which range from the grasslike shrub *Haloxylon ammodendron* to the gigantic desert poplar *Populus euphratica*. Then, they are awarded a certificate attesting to their tree-planting deeds. Through GPS visualization, users can check on images of actual trees they have asked Alibaba to plant (see figure C.3). Throughout this process, in its motivation system, *AF* operates as an addictive computer game, but it markets itself as a promising and benevolent project because of its connection to real-world "rewards." One user reports that he has to get up at 6:30 every morning to collect the forest energy. A parent reports that his four-year-old child changed his daily habits to wake up early to reap more forest energy after he began to use *AF*.[6] While the "reward" is abstract for the city users in that they will not actually experience these trees in an immediate sense, the app provides a kind of virtue signaling and simulated tree access. This simulated tree access effaces the real distance between most users and the real trees.

The design of *AF* accomplishes two essential aspects of digital communication—affect and expertise—by evoking the sympathy of users and lowering the level of expertise associated with tree planting. In *Affective Publics: Sentiment, Technology, and Politics*, Zizi Papacharissi argues that "affect presents a key part of how people internalize and act on everyday experiences."[7] Papacharissi claims that the cognitive and the affective are able to empower individuals to change their actions. Thus, when users see the benefits of trees through *AF*, they are likely to internalize them and perform environmentally friendly actions in their daily lives. One user explains that in order to accumulate forest energy via the app, he walks to work rather than traveling by bus, while other users pay their utility bills via their cellphones instead of driving to pay in person.[8] As a result of these actions, less automotive exhaust is produced. Furthermore, users who lack the time, knowledge, access to nature, or ability to conduct physical labor can still plant trees in a desert. Detaching real practices from environmental expertise thus does not hamper but instead contributes to afforestation because it lowers the barriers to tree planting. As a result, more and more individuals become eager to use *AF*.

As "Songs of the Forest," discussed in chapter 4, employs sound to cul-

FIGURE C.3 **Images of trees planted in Inner Mongolia in August 2019. Source: Alibaba, *Ant Forest*.**

tivate a new soldier, *AF* invites users to watch a short film clip with audio to "hear" nature when they first activate their accounts. The film, showing only white words against a black background, invites users to "imagine" afforestation purely through sound. The appealing soundtrack of rain, creeks, buzzing bees, and black-necked cranes flying over a conservation zone evokes a sense of tranquility and marks a departure from bustling urban life. Using these sounds and words, the film invites users to imagine an Edenic world where forests have been regenerated. The film shows that tree planting is not merely about trees but also about creating a fully affective landscape—a lively, charming, and promising ecological system.

AF is not only an affective text but also an interactive platform that socializes its users.[9] The app's interface glorifies those who earn the most grams through their daily activities. It ranks the participants, spurring them to compete for more grams. *AF* is not purely competitive, however; it also encourages collaborative behaviors and group projects. A user may, for instance, send a boosting tool to help another user's small tree to "grow" faster. Users from a designated school or university can cooperate and contribute to tree planting together. Users plant trees with their family members, lovers, classmates, colleagues, and friends. As a result, *AF* builds a society of individuals in solidarity with each other because it calls all of its members to focus on the same goal: tree planting. In so doing, they exchange green behaviors not so much for trees or conservation per se but for a sense of moral validation as well as of belonging to a group. As a result, *AF* is not just simulating communities but creating an imaginary nation on the platform.

When users of different social classes participate in tree planting, they may actually help generate a more equal society. Whereas the Republican era's Arbor Day ceremonies manifested and reinforced politicians' leadership, *AF* instead attempts to establish an equal society where every participant assumes the same power—which may even prove to be more effective than a single figure's domination. As the name *Ant Forest* suggests, a small group of ants can band together and generate massive collective power. *AF* calls attention to what Jing Wang calls *incrementalism,* in which Chinese civil society seeks a nonconfrontational means to promote the activist agenda of the masses.[10] In calling up this association with collective power, the brand name suggests that ordinary individuals, not just the affluent, can contribute to afforestation and environmental conservation. *AF* therefore becomes a powerful and feasible means of fighting against what Rob Nixon calls *slow violence*—environmental degradation whose "delayed destruction . . . is dispersed across time and space"—in regions suffering from desertification, and the app can thereby help improve the lives of the poor.[11] As a result, *AF* generates added value by promoting a nonhierarchical society.

In addition to reducing the gap between social classes, *AF*'s tree planting also bridges the rural-urban divide that plagues contemporary Chinese society. *AF* demonstrates that reducing carbon dioxide emissions in the city can contribute in tangible ways to afforestation in the desert. This is, in fact, a reprise of the Mao-era practice in which urban residents would go to the countryside and improve it by planting trees. In the case of *AF*, city users do not actually have to trek into the desert—they just purchase items, and Alibaba hires workers to plant trees on their behalf. In this regard, this process of tree planting by *AF* is an improvement upon the highly intrusive practices showcased in chapter 5 in which the educated youth from the city imposed their conceptions of nature on the residents in the countryside. Instead, by mediating between the rural and the urban, *AF* encourages a productive interplay between these two regions that are usually highly separated in Chinese society.

As an environmental initiative, therefore, *AF* functions simultaneously as a consumer portal, a cellphone game, a social network, and a charity project. It calls for the active participation of different social classes, confounds the boundary between the virtual and the real, and narrows the gap between the city and the countryside. *AF* embodies fluid and adaptative dimensions of Chinese conservationist ideas as it stands as a premier example of how tree planting can create social equality and, at the same time, promote active citizenship, implicitly imbuing Chinese citizens with warm and caring sentiments.

It is illuminating to compare *AF*—the virtual Arbor Day practice—with the original Arbor Day of the early twentieth century. They both attempt to advance the fundamental roles of trees in fighting against ecological woes. In teaching individuals that there is an urgent need to address ecological problems, Arbor Day and *AF* cultivate responsible citizens through tree planting. While Arbor Day encourages participants to plant trees and then keep environmental conservation in mind, *AF* reverses the process by encouraging users first to keep environmental conservation in mind and then to plant trees. While Arbor Day was originally conceived as a one-day celebration, *AF* makes it possible for Arbor Day to take place every day. And although Arbor Day has ultimately failed to achieve Sun Yat-sen's goal of shaping Chinese minds with regard to environmental conservation and became a symbolic gesture among the Nationalists, *AF* penetrates into its users' daily lives, encouraging and rewarding their conservation efforts.

Consumer Environmentalism: Reconciling Conservation and Destruction; Negotiating Socialism and Capitalism

Despite its promise to afforest China, *AF*, at its most fundamental level, is a profit-seeking enterprise. In promoting its afforestation project, *AF* encourages its users to subscribe to services offered by Alibaba, giving the firm an advantage over its main competitor, Tencent. One user explains that he could pay his bills through either Alibaba or Tencent; when he realized that he could contribute to afforestation by using Alibaba, he decided on Alibaba.[12] *AF* invites its users to patronize Alibaba's businesses, including Alibaba's lending company, food delivery, and online banking, all of which have become closely associated with *AF*. When one gets a loan from Alibaba or buys food online, one receives grams and contributes to tree planting.

This somewhat unusual mode of business is still business, however. Consumer capitalism, even if offset by tree planting, contributes to further environmental degradation. In the reusable-cup example mentioned above, the promotion accepted by *AF* can be used in Starbucks in only two cities: Beijing and Shanghai. And only a small portion of Starbucks products participate in the *AF* project. As consumers acquire a more positive image of Starbucks, they tend to shop there more. As its sales increase, its use of non-environmentally friendly products—which account for a major part of its revenue—expands.[13] As a result, Starbucks's old mode of business—which actually harms the natural world—persists.[14] Consumers, however, are rendered complacent by their largely false perception that they have purchased products from an environmentally friendly company.

In this regard, *AF* is an example of consumer environmentalism more broadly, promoting environmental conservation while at the same time advancing a consumer culture that is controlled by the affluent and by megacorporations with little concern for anything besides profits.[15] Peter Dauvergne argues that "one effect of the increasing dominance of environmentalism of the rich is to mute calls for an overhaul of consumer capitalism even as the global environmental crisis escalates."[16] He complains that "big companies deploy the language of sustainability to obfuscate, and in some instances even promote, unsustainable and inequitable consumption."[17] Dauvergne sheds light on the game Alibaba is playing. Camouflaged as an environmentally friendly company, e-commerce behemoth Alibaba skirts criticism by pretending to promote an eco-friendly business. As a result, pseudoenvironmentalism itself becomes a business. When conservationist ideas circulate among Alibaba's consumers, the firm increases its economic value by manipulating the presumed universal positive benefits of trees.

AF illustrates the paradox and nuances of environmentalism in contemporary China as it uses Communist rhetoric that stresses collective effort to promote a capitalist goal of making a profit. Anthropologist Lisa Rafel argues that while "socialist power operated on the terrain of 'consciousness,' post-socialist power operates on the site of 'desire.'"[18] *AF* integrates both consciousness of political mobilization, on the one hand, and desire generated by neoliberal public culture, on the other hand. On the surface, *AF* stresses individualist contributions to environmental conservation. Yet the glorification of collective contribution over individualistic effort—which was advocated in the Mao era—persists. *AF*'s Communist-influenced rhetoric—that a small figure can contribute to a large project—is particularly redolent of Mao-era propaganda examined in chapters 3 and 4 in which even a simple Mongolian farmer or a young soldier can contribute to a grand historical goal. Exploring the Mao-era cultural politics between the national agenda and individual reactions, Ban Wang argues that in China's discursive practice, the political became an aesthetic experience: politics "has been made to take on the intensity, passion, pleasure, and pain of the individual's lived experience," producing visual and auditory joy in socialist China.[19] To some extent, *AF* employs visual and auditory elements to advance a politically populist and collective goal based on the notion that individuals, when joined together, can make a difference as a large crowd.[20] As a project of the Ant Financial Services Group of Alibaba, however, *AF* operates as a means to expand the firm's business to make a profit. As we have seen, the success of afforestation depends on its users' continuing interactions with Alibaba's businesses, ranging from ordering fast food to taking an Uber to making payments online. For all its arboreal imagery, the kernel of *AF* is a moneymaking machine.

Nevertheless, *AF* has received accolades from its users. On the Quora-like Zhihu, a popular Chinese question-and-answer forum, the most popular response to the question of whether Alipay's *AF* really helps ecology or is just a superficial effort is that *AF* epitomizes the success of corporate contributions to public welfare. While users know that Alibaba is expanding its own business by promoting *AF*, most of them still believe that it contributes to the welfare of the nation. One user argues, "We know that *Ant Forest* is for a businessman to make money. But when he is making a profit from Alibaba, we should praise Ma Yun's kindness to advance public welfare instead of criticizing its greed."[21] Indeed, *AF* has gained currency even among the poor because it channels their desire to create a kind and environmentally friendly society. One user in a small town who earns only 3,000 yuan (approximately 400 US dollars) a month contributes twenty yuan (approximately three US dollars) a month to plant trees in the desert, even at times when she can barely make ends meet.[22] She has faith in *AF* because she believes it serves as a charity project that will benefit future generations.

The reception of *AF* challenges our understanding of the app as a simplistic binary opposition between making money and promoting social welfare, thereby revealing the nuances of Chinese environmentalism, which encompass disparate and even contradictory impulses. It shows how a moneymaking machine can portray itself as an environmentally friendly enterprise and receive widespread appreciation from the public. In this process, the success of afforestation in the desert, the encouragement of consumerism in urban regions, and the promotion of conservationist environmental ideas in both places all occur at the same time. The simultaneous display of multiple and even contradictory attitudes demonstrates how an understanding of Chinese environmentalism must not only take a specific time and place into consideration but also the individuals who put conservationist notions into practice. While *AF*'s users increase their purchases of items, they also protect the environment by conducting more environmentally friendly activities or educating their children on the need to conserve nature.

In the case of *AF*, consumer environmentalism manifests and integrates all dimensions of Chinese conservationist ideas examined in this book: universality, fluidity, and adaptation. The conception that tree planting provides *universal* benefits offers Alibaba an edge in advertising itself as a responsible company in China and in the wider world. In advancing its afforestation project, Alibaba spurs Chinese citizens to participate in environmental conservation and transmit environmental ideas in an interactive and dynamic way. Users are constantly rewarded for their actions, building a virtuous cycle of dynamic engagement between users and Alibaba. *AF* also operates as a social lubricant as the knowledge of tree planting flows from the city to the countryside. At the heart of its *fluid* endeavors in transmitting environmental ideas lies Alibaba's goal of commodifying conservation by transforming green ideology into social goods and then into profit. This process demonstrates how *adaptive* environmental rhetoric is. Employing traditional socialist rhetoric of bottom-up and collective interests, *AF* actually implements an e-commerce giant's ambition to sustain and expand its business from the top down. In this postsocialist era when socialism is integrated with capitalism, the socialist rhetoric of grassroots political participation merges with the pursuit of money-driven goals. As a result, environmental rhetoric is always changing itself, going beyond a narrow definition of environmental conservation and manifesting its ability to survive and thrive in various social and political conditions. In this process, environmental ideas accommodate contested notions, including socialist collective mobilization and capitalist moneymaking desire and the need to encourage potentially eco-unfriendly consumption and the need to inculcate conservationist ideas. These tensions demonstrate the vitality as well as the contingency of Chinese environmental attitudes.

Because of these features, Chinese conservationist ideas become what might be called *green currency*. A corporation "deposits" trees into the "bank" of conservation and later "withdraws" much more: a consumer culture, a positive image and social trust, a strengthened social network, a tool to cultivate its users, and a viable method of fighting desertification and global warming. The dividends paid by trees thus far exceed the principal invested. This accruing interest further promotes positive images of trees, which in turn compound the interest, creating a virtuous environmental cycle. *AF* thus serves as a textbook case for mixed environmentalism in China. Integrating the profit-driven dimension of consumer environmentalism and the goal of creating active citizenship and an egalitarian society, *AF* constitutes an intricate and nuanced lens through which to understand modern Chinese culture. From Arbor Day in 1916 to virtual Arbor Day in 2016, from the Nationalists to the Communists, from foresters to politicians, from the borderland to the heartland, from offline to online, Chinese conservationist ideas circulate among the people and convey a notion that is much more ambitious than environmental conservation itself. A saga of trees strikes a chord with most of the Chinese citizenry and reflects the environmental ideas of one of the most populous countries on Earth.

Do Trees Have a Nationality? A Global Afforestation Saga

AF is dedicated to participating in an energy revolution by reducing global carbon dioxide emissions from the grass roots. Efforts to curb greenhouse gases have gained momentum in the past decades. Signed in 1997 and becoming international law in 2005, the Kyoto Protocol aimed to reduce greenhouse-gas emissions by 5.2 percent by 2012. Countries were assigned maximum carbon-emission levels for certain periods and participated in carbon-credit trading. If a nation emitted more than its assigned limit, its emissions limit would be lowered in the following period. The Kyoto Protocol's efforts were continued in the 2016 Paris Agreement, which aims at curtailing greenhouse-gas emissions so that average temperature increases remain within 2°C above preindustrial levels, significantly reducing the risks of climate change.[23] *AF* functions as the world's largest platform for personal carbon accounts, which record each individual's low-carbon behavior in quantitative terms. Eric Jing, CEO of Ant Financial, claims that *AF* "is essential to complement top-down action, such as the Paris Agreement and the 2030 Agenda on Sustainable Development."[24] Planting one million trees will reduce carbon emissions by around eight million tons a year. These trees are becoming a considerable green asset in the global carbon market for Alibaba and even for China, the largest greenhouse-gas emitter in the world as a whole. When conservation motives are deposited into the "world bank" of conservation to become green currency, they withdraw bottom-up political engagement, a forward-looking attitude, and an awareness of global citizenship.

While the Chinese central government's afforestation project has achieved only limited success in controlling desertification, *AF* shows the vitality of civil society in managing the natural world and engaging with local governments. Because of limited investment and poor management, the Three-North Shelter Forest program, also known as the Great Green Wall of China, started in 1978 and targeting a large part of northern China, has not been effective in reversing desertification. Even where there have been successes, their impacts are largely unknown to the public.[25]

AF has proved to be a more effective alternative. Spearheaded by China's then-richest businessman, Jack Ma, it has attracted millions of users who consciously or unconsciously contribute to afforestation in their everyday lives. On the rural grassroots level, the money from *AF* has funded village-level local governments—which are usually extremely short on funds in isolated regions—to hire more locals to plant trees.

By targeting grassroots participation from both the rural regions and cities, *AF* serves as a wedge of authoritarian environmentalism, which is usually top-down and nonparticipatory. In *Carbon Democracy,* Timothy Mitchell argues that the energy revolution gave rise to participatory politics at the grass roots because coal-mine production enabled miners to organize and assume considerable power.[26] By transforming tree planting into a virtual process based on individuals' daily actions, business giant Alibaba empowers its users to make real changes and encourages them to involve themselves in environmental initiatives. With more than five million users and active cooperation with local governments, business giant Alibaba has shown the adaptability of Chinese conservational ideas by meeting the local governments' needs for afforestation and answering the central government's call for promoting environmental notions among the general public.

AF, though tainted by its impetus to make a profit, nevertheless is afforesting China and contributing to a dynamic engagement between the rural and urban, upper and lower classes, and China and the world. According to developers of *AF,* if every person would participate in tree planting in China, witnessing the growth of trees and enjoying the rewards of their everyday practices, China would better position itself globally, and the Chinese people would become stronger and healthier, connecting to each other through spontaneous practices of caring, affection, and respect. This goal is exactly what Chinese foresters were so genuinely and passionately aspiring to a century ago.

Conclusion

Trees—as an industrial product, a financial investment, an ecological weapon against flood and drought, and a vehicle to trigger nationalistic sentiment—have shaped Chinese society comprehensively and deeply. The tension between universality and fluidity gives rise to a new evolution of Chinese conservationist ideas. The purported universal value of trees and forests has motivated Chinese foresters, politicians, and writers to promote afforestation efforts. As these ideas circulate across time and space, they encounter resistance and then adapt to changing political and societal circumstances. In this process, novel environmental ideas arise, but older ones do not disappear, eventually producing conflict among various forms of environmentalism.

Modern Chinese literature and culture demonstrate the contestation of environmental ideas and the ways these ideas have accommodated various contexts. Western-educated foresters promoted forestry as an academic discipline as well as an enterprise capable of boosting the Chinese economy, society, and welfare, eventually leading to the establishment of Arbor Day. While the Nationalist politicians deployed Arbor Day as a symbolic gesture to justify their political legitimacy, the Communists promoted tree planting in Shaanxi Province without making reference to Arbor Day. After the Communists came into power in 1949, they extended this tree-planting initiative nationwide. The foresters' vision of nature as a dynamic system on a broad scale was at odds with the hydrologists' more static, isolated, and small-scale perspective. While the opinions of the hydrologists predominated at first, the two views eventually became integrated with each other, demonstrating the flexibility and resilience of conservationist ideas. When these purportedly universal afforestation ideas arrived in the northern borderland, however, they encountered local resistance. The ethnic minority's nomadic understanding of the natural world ran against the settled vision of nature embraced by the Han Chinese. The contestation of these different views engendered fierce conflict. This confrontation may escalate in a more combative manner in the forests: the Communists imposed a militaristic understanding of the natural world in place of a peaceful and romantic interpretation. They spared no efforts to eliminate those who hid in the forest and evaded state power. After this goal was accomplished, the desire to use the resources of the forest came to the fore, causing serious deforestation and natural calamity. It was not until the 1980s that explicit conservationist environmental literature formally began to appear to document these consequences. The trend of adapting Western conservationist ideas to the Chinese context stood in contrast to a nativist strategy of excavating Chinese traditions to combat deforestation, but neither trend appealed to the population at large. Instead, it is the cellphone app *AF* that has activated five hundred million users to participate in afforestation projects. While it is, at the base, a lucrative capitalist enterprise, *AF* incorporates traditional, socialist, grassroots-mobilization rhetoric to promote afforestation in rural regions.

Contested Environmentalisms has endeavored to demonstrate that Chinese conservationist ideas perpetually involve ongoing negotiation between multiple elements, resources, and constraints. Rather than a state of being, Chinese environmentalisms are always in the process of becoming. They actively integrate novel and even contradictory modes of thought, reflecting a heightened degree of contingency and openness. Chinese environmentalisms, as embodied in the saga of trees and forests, manifest multiple directions that may confound and confuse the original intentions of afforestation. While environmental ideas adapted to the new social and cultural environment, they added other goals and compromised their originally singular and sharply defined objectives of promoting conservation. Chinese conservationist ideas are affected and even tainted by their byproducts, such as social mobilization, cultural dominance, and political legitimacy.

By using trees as an anchor, this book has shown the multivalent character of Chinese environmental initiatives. When one "deposits" trees into the "bank" of conservation, one may eventually withdraw a fertile cultural currency with as many positive valences as the newspaper list of slogans reproduced in the introduction: an index of progressive culture, a positive image of caring, a desire for social trust, a strengthened social network, a tool to cultivate youth, a method to help a nation navigate the future, and an attitude of world citizenship. The belief in the benefits of trees that circulates throughout China operates as a historical force that implies considerably more than the natural environment itself.

Using China as a case study, this book tries to answer a fundamental and deceptively simple question: What *is* the environment? By not treating the environment either as uniform or given, this book circumvents the reification of modern environmentalism into an ontological category, a state of being, or a teleological destination that has or has not yet come to fruition in any given case. The multiple forms of Chinese environmental ideas (symbolic vs. real; static vs. dynamic; romantic vs. militarized; native vs. foreign; capitalist vs. socialist) make the answer fundamentally contingent. While foresters, writers, and cultural workers promote their conservationist agendas, they are constrained by their modes of training and knowledge systems. We have seen how foresters' and hydrologists' different ways of defining the environment shape their conservation rhetorics and practices. Similarly, the Han Chinese and Mongolians' contrasting understandings of the desert and the natural environment at large compels us to reconsider the limits of promoting universal conservation. These observations are not confined to Chinese environmental studies. As noted in the introduction, the prominent US marine biologist and environmentalist Rachel Carson's 1962 *Silent Spring*, showing the dangers of the pesticide DDT, triggered not only backlash from chemical industries but also issues among environmental scientists themselves. Animal advocates and soil scientists celebrated her endeavor because ending the use of pesticides protected animals and soil. For public-health experts, however, pesticides were indispensable in killing the vector of malaria and preventing famine, especially in the global South. For US foresters, stopping the use of DDT meant certain damage to the health of forests with the return of the gypsy moth and Japanese beetles.[27] Defining what precisely counts as the *environment* varies not only with each group's interests and training but also with the cultural parameters that define and refine the natural world.

In the general field of environmental humanities studies, this book contends that conservationist ideas are historical forces that flow and thereby change across time and space. Modern Chinese conservationist ideas have acquired their value and have assumed their power not because their meanings are universal or solid but precisely because they are so malleable, adaptive, and resilient when they encounter new and even unexpected circumstances. We therefore need an integrated cultural, historical, scientific, and literary analysis to understand this dynamic process in the fields of environmental humanities and Chinese studies. Such an inclusive approach will allow us to explore the underlying ideas that drive Chinese environmental writings and practices.

As discussed in the introduction, this book has set out to make three key contributions to both China studies and environmental humanities. First, it uncovers how forests mold subjecthood and serve as an essential part of Chinese nationhood. Second, by using the ideas of trees and forests as an entry point into modern Chinese environmental ideas, the book demonstrates various ways in which modern Chinese environmentalism is contested. Third, *Contested Environmentalisms* offers a cultural and historical perspective on Chinese authoritarian environmentalism. I would like to conclude by further reflecting on these three points.

First, *Contested Environmentalisms* demonstrates how trees mold nationhood and subjecthood in a multivalent, fluid, and contingent manner. Trees and forests deeply shape almost all aspects of life, significantly affecting political mobilization and legitimacy (chapter 1), scientific reasoning (chapter 2), ethnic identity (chapter 3), youth growth (chapter 4), military mobilization (chapter 4), intellectual reflection (chapter 5), and the technological enterprise (coda). In many ways, the Chinese revolution is a forest revolution, a revolution of disseminating conservationist ideas. In addition to contributing to real tree planting, the forestry revolution also acquires temporal and metaphorical dimensions: the growth of a tree coincides with that of the youth and the nation, in many ways fitting the genre of the bildungsroman. From the New Life movement led by Chiang Kai-shek through the Greening the Motherland Campaign in the Mao era to the young soldier in "Songs of the Forest," trees embody optimistic, lively, and forward-looking values and instill these cultural values across time and space.

Uncovering Chinese environmentalisms also encourages us to reconsider the different forces operating in Chinese politics and culture. These forces become involved with, compete against, and even cancel out each other, calling on us to evaluate Chinese environmental notions in a more impartial manner. On the one hand, we see the scientific and rational aspects of Chinese environmental ideas. In the Mao era, China had sufficient flexibility to integrate scientific notions to advance the agenda to fight against desertification, soil erosion, and flooding on a national scale. On the other hand, we have seen the difficulty of getting the local and Indigenous voices to engage with and even oppose the one promoted by the state. It is this contradiction that drives environmentalism in modern China.

Second, analyzing the cultural and historical dimensions of Chinese authoritarian environmentalism provides a mirror for understanding the resilience of Chinese authoritarian regimes. While authoritarian China is largely oppressive in nature, it does include a variety of voices, at least in science. The Chinese government deploys and integrates scientific views in its governance. Going beyond the political-ideology divide, the early Mao-era Communist government embraced ideas from capitalist countries (chapter 2). While the strongman Mao dominated China for decades, the post-Mao government has featured cultural flexibility to tolerate various voices. The presence of two distinctive directions in the 1980s (chapter 5) illustrates that authoritarian China was inclusive and resilient enough to recover from its economic, cultural, and ecological woes. After the 1990s, China steered itself into the international community when it combined economic reform with continued tight political control. China has demonstrated its flexibility by evoking two contentious rhetorical fields: individualist, capitalist moneymaking and socialist mass mobilization (coda).

Instead of treating it as a fatal flaw, this book argues that contestation contributes to authoritarian China's strength. Using environmental attitudes as a case study, it contends that contestation constitutes an inherent feature of the Chinese cultural repertoire and social fabric: these attitudes are constantly negotiating with each other but are not ultimately capable of being resolved. Chinese environmental contestation shows that Chinese society can evoke various forces, and even opposing ones, to enlist a broader audience. Whenever needed, Chinese environmentalism can slide to one end or the other of the spectrum, empowering and energizing itself. Despite the regime's determined promotion of a single version of Chinese environmentalism, it is exactly the fact that it is not uniform that enables Chinese governance to grapple with various, ever-changing, and precarious scenarios. Chinese environmentalisms may not speak with a dominant voice, but they gain breadth by inspiring a vast range of diverse individuals to disseminate multiple variations of conservational attitudes across China.

Contestation functions as a safety valve, providing a cultural platform from which to consider opinions from various backgrounds and contributing to the vitality of Chinese society. It also forges more stable and responsive governance. By including disparate voices, contestation offers various parties an arena in which to compete, negotiate, and reconcile, thereby reducing the destructive potential of radical thoughts and decisions. Recent history shows that overlooking these voices may engender devastating consequences. Prioritizing a single dominant opinion, Xi Jinping's regime downplayed and eliminated disparate voices both culturally and politically, triggering drastic social issues. "A healthy society should not have just one voice," said the late Li Wenliang, the COVID-19 whistleblower in Wuhan.[28] His early warning of the new virus led only to his being ordered by the police to remain silent. Forced to acknowledge the single voice of state power, Li saw his professional voice overlooked, and China lost precious time for grappling with the coronavirus just as it began to be widely observed in Wuhan. This story demonstrates how Chinese society, and even the entire world, can suffer drastically when contradictions are eliminated to present a seemingly immaculate society. We are now in the Mao 2.0 period, an era that has become increasingly authoritarian, and the contested environmentalisms examined in this book are all the more relevant and illuminating in this context. The symbolic, scientific, ethnic, militaristic, intellectual, and economic frictions elaborated in each chapter of this book offer strategies for effective policymaking in a fluid and tumultuous geopolitical context. Probing the multiple dimensions of contestations not only offers a guide to understanding Chinese environmental attitudes but also provides a prescription for engaging with an increasingly oppressive regime.

Third, the historical dimension of Chinese authoritarian environmentalism casts light on the cultural politics and practices of contemporary China, which offers insight into policymaking. It offers us a historical trajectory of how Chinese environmental issues have advanced to their current stage. From Chiang Kai-shek's New Life movement—which encouraged tree planting as a way to instill order and discipline in the Republican era—to the Mao era's Greening the Motherland Campaign to building the Great Green Wall in the post-Mao period, authoritarian China has organized society through mass mobilization and forced individuals to align with the state's goals. These historical moments provide a fertile cultural repertoire to be evoked, revitalized, and reinvented.

History does repeat itself. When Xi Jinping came to power, he reinvented many Mao-era environmental rhetorics and programs. Like the Mao era (chapter 2), Xi's government also integrates various scientific voices, for example, to endorse the establishment of national park systems. The idea of establishing a national park also evokes the symbolism of environmental conservation (chapter 1) in order to present China as being as environmentally friendly as Western countries, especially the United States. The construction of these national parks comes, however, at the expense of the local voices when the residents are forced to accept this exercise of state power (chapter 3).[29] The history of modern Chinese environmentalism is by no means purely about the past as China recycles its cultural practices and ideologies examined in this book and reactivates them in the present and into the future.

Probing the cultural and historical contestation of Chinese authoritarian environmentalism indicates China's strategic efforts to align itself with the international community while simultaneously pursuing its own expansive ambitions.[30] On the global stage, promoting many environmental initiatives through its Belt and Road Initiatives (BRI), China is demonstrating its ambition to become a world leader and a global empire. Challenging the US-dominated world order and pushing back against the much-touted US "pivot to Asia," BRI aims to promote infrastructure connectivity across Eurasia and east Africa and expand China's influence worldwide.[31] Moreover, China has announced and affirmed its ambition to become carbon neutral by 2060, a decision that won international acclaim at a time when the United States exited the Paris Climate Agreement during the Trump administration. China's public embrace of conservationist attitudes became a strategy to expand the emerging Chinese empire. China has touted its tree-planting efforts in order to brandish its achievements in controlling climate change, while at the same time expanding its carbon footprint overseas through the BRI.[32] The pursuit of a self-interested political agenda can and does coincide with the advancement of global welfare. The tension between promoting a seemingly altruistic environmental agenda and pursuing aggressive political ambitions necessitates that environmental practices adapt to the vastly different political climates of the world's two leading powers. This adaptation will require a fresh round of negotiations and reconciliation among various domestic and international stakeholders. Regardless of what this process entails, the nation that takes the lead in addressing the urgent environmental challenges faced by all of humanity will likely be recognized as the true leader of our planet. That is China's goal.

Despite their apparent incoherence, Chinese conservationist attitudes hold immense transformative power as they drive China's green rise. Over a century ago, Western-educated Chinese foresters promoted the benefits of trees, which sparked a convoluted journey filled with twists, turns, unity, division, and contingencies. Constant contestation has paved the way for innovation, adaptation, and resilience, making Chinese conservationist attitudes a potent force for change. Paradoxically, the strength of Chinese environmentalisms lies in an ostensible weakness: their inconsistency and contestation, which usher in a new order.

NOTES

Introduction

1. Trees are associated with the rise of modern environmentalism. Richard Grove, *Green Imperialism: Colonial Expansion, Tropical Island Edens and the Origins of Environmentalism, 1600–1860* (New York, NY: Cambridge University Press, 1995); Gregory Allen Barton, *Empire Forestry and the Origins of Environmentalism* (New York, NY: Cambridge University Press, 2002).

2. "Zhishu biaoyu" 植树标语 [Tree-planting slogan], special issue, *Xinwen bao* 新闻报, March 12, 1928. My translation. All translations in this book are mine unless otherwise stated.

3. Tree cutting does not necessarily harm the natural environment. In fact, it can promote forest growth when done properly. The real issue arises from excessive and destructive resource extraction. Here, I am highlighting the potential threats posed by tree cutting.

4. As environmental historian William Cronon contends, "In narrating those consequences, we inevitably interpret their meaning according to human values—but the consequences themselves are as much nature's choice as our own. To just that extent, nature coauthors our stories." Recent studies in biology and environmental sciences indicate that plants can see, smell, and feel. They can also broadcast their responses to other plants, insects, and even people. The resilience of conservationist ideas stems not only from environmental scientists and writers who adjust narratives to fit various purposes but also from trees, whose charismatic appeal and beneficial values undergird and "coauthor" the political and social twist of conservation rhetoric. William Cronon, "A Place for Stories: Nature, History, and Narrative," *Journal of American History* 78, no. 4 (1992): 1373. See also Michael Pollan, *The Botany of Desire: A Plant's-Eye View of the World* (New York, NY:

Random House, 2001); Daniel Chamovitz, *What a Plant Knows: A Field Guide to the Senses* (New York, NY: Farrar, Strauss and Giroux, 2012).

5. Rachel Carson's *Silent Spring* (1962), with its revelation of the danger of the pesticide DDT to the environment, expresses the need to evaluate environmental values in their contexts. For some, the book was an act of heroism; for others, it was an irresponsible breach of scientific objectivity. But the story is more complicated than a binary opposition between environmentalists and the chemical industry. The book triggered disparate reactions even among environmental scientists. Animal advocates and soil-science groups celebrated Carson's endeavor because ending the use of the pesticide protects animals and soil. For medical experts, however, pesticides were indispensable in killing the vector of malaria—the mosquito of the genus *Anopheles*—and also helped prevent famines caused by crop-destroying insects. For foresters, stopping the use of DDT meant certain damage to the health of forests through the return of the gypsy moth and Japanese beetle. Each side represents its own interests and reflects its training. Rachel Carson Papers, box 83, Beinecke Rare Book and Manuscript Library, Yale University. For studies on Rachel Carson, see Mark Hamilton Lytle, *The Gentle Subversive: Rachel Carson, Silent Spring, and the Rise of the Environmental Movement* (New York, NY: Oxford University Press, 2007); Lisa H. Sideris and Kathleen Dean Moore, eds., *Rachel Carson: Legacy and Challenge* (Albany: State University of New York Press, 2008).

6. Environmental social movements in a global, postcolonial context with far-reaching political implications illuminate China's own problems. In the postcolonial context, planting trees operates as a catalyst for social movements. In India, the Chipko movement in the 1970s that aimed at protecting trees also became a women's movement, a peasant insurrection, and a prodevelopment protest. In Kenya, Nobel Prize winner Wangari Maathai heralded the Green Belt movement against the ruling dictatorship in the 1980s by planting trees. Rob Nixon, *Slow Violence and the Environmentalism of the Poor* (Cambridge, MA: Harvard University Press, 2011), 128–49; Cajetan Iheka, *Naturalizing Africa: Ecological Violence, Agency, and Postcolonial Resistance in African Literature* (New York, NY: Cambridge University Press, 2018), 126–37. For a depiction of the Chipko movement, see Anna Tsing, *Friction: An Ethnography of Global Connection* (Princeton, NJ: Princeton University Press, 2005), 213–44.

7. Dai Qing 戴晴, ed., *Changjiang, Changjiang: Sanxia gongcheng lunzheng* 长江, 长江: 三峡工程论争 [Yangtze, Yangtze: the Three Gorges project debate] (Guiyang: Guizhou renmin chubanshe, 1989). See also Elizabeth Economy, *The River Runs Black: The Environmental Challenge to China's Future* (Ithaca, NY: Cornell University Press, 2011), 142–45.

8. For modern environmental notions, see Paul Warde, Libby Robin, and Sörlin

Sverker, *The Environment: A History of the Idea* (Baltimore: Johns Hopkins University Press, 2018); Roderick Nash, *Wilderness and the American Mind* (1967; repr., New Haven, CT: Yale University Press, 2014); Jedediah Purdy, *After Nature: A Politics for the Anthropocene* (Cambridge, MA: Harvard University Press, 2015); Grove, *Green Imperialism*.

9. Recent discussions of the Plantationocene reconsider large-scale global environmental alternations from the perspective of small-scale local experience. Donna Haraway, e.g., contends that the emergence of plantations disrupts local natural environments and fundamentally simplifies local species. The promotion of monoculture was often exercised in a repressive and hegemonic manner. In the United States, the use of slave labor in the plantation economy led to ecological simplifications as well as the large-scale abduction and enslavement of Africans. In China, monoculture timber plantations and agricultural development contextualize the Plantationocene. On the one hand, the government's commitment to afforestation in China's borderland as a way of fostering Chinese nationalist sentiment ultimately has its origins in the West; on the other hand, when this monocultural afforestation comes to embody hegemonic state power in shaping the local natural environment, forestry conveys dominant Han Chinese values as forestry is demoted to another form of agriculture. Donna Haraway, "Anthropocene, Capitalocene, Plantationocene, Chthulucene: Making Kin," *Environmental Humanities* 6, no. 1 (2015): 159–65; Gregg Mitman, "Reflections on the Plantationocene: A Conversation with Donna Haraway and Anna Tsing," Edge Effects, last modified October 12, 2019, https://edgeeffects.net/haraway-tsing-plantationocene/; Yunan Xu, "Land Grabbing by Villagers? Insights from Intimate Land Grabbing in the Rise of Industrial Tree Plantation Sector in Guangxi, China," *Geoforum* 96, no. 4 (2018): 141–49.

10. For the colonial nature of environmental practices, see Ramachandra Guha, "Forestry in British and Post-British India: A Historical Analysis," *Economic and Political Weekly* 18, no. 44 (1983): 1882–96; Shaul E. Cohen, *Planting Nature: Trees and the Manipulation of Environmental Stewardship in America* (Berkeley: University of California Press, 2004); Diana Davis, *Resurrecting the Granary of Rome: Environmental History and French Colonial Expansion in North Africa* (Athens: Ohio University Press, 2007); Richard Widick, *Trouble in the Forest: California's Redwood Timber Wars* (Minneapolis: University of Minnesota Press, 2009); Joachim Radkau, *Nature and Power: A Global History of the Environment*, trans. Thomas Dunlap (New York, NY: Cambridge University Press, 2009); Pamela McElwee, *Forests Are Gold: Trees, People, and Environmental Rule in Vietnam* (Seattle: University of Washington Press, 2016); David Fedman, *Seeds of Control: Japan's Empire of Forestry in Colonial Korea* (Seattle: University of Washington Press, 2020).

11. Raymond Williams, *Marxism and Literature* (New York, NY: Oxford University Press, 1978), 125.

12. R. Williams, *Marxism and Literature*, 125.

13. Karen Thornber, *Ecoambiguity: Environmental Crises and East Asian Literatures* (Ann Arbor: University of Michigan Press, 2012); Ban Wang, *At Home in Nature: Technology, Labor, and Critical Ecology in Modern China* (Durham, NC: Duke University Press, 2023); Robin Visser, *Questioning Borders: Ecoliteratures of China and Taiwan* (New York, NY: Columbia University Press, 2023); Corey Byrnes, *Fixing Landscape: A Techno-Poetic History of China's Three Gorges* (New York, NY: Columbia University Press, 2019).

14. B. Wang, *At Home in Nature*; Visser, *Questioning Borders*.

15. Byrnes, *Fixing Landscape*. For Chinese-language scholarship on ecocriticism, see my early overview article, Cheng Li, "Echoes from the Opposite Shore: Chinese Ecocritical Studies as a Transpacific Dialogue Delayed," *Interdisciplinary Studies in Literature and Environment* 21, no. 4 (2014): 821–43. For some recent edited volumes, see Sheldon Lu and Haomin Gong, eds., *Ecology and Chinese-Language Cinema: Reimagining a Field* (New York, NY: Routledge, 2019); Chia-ju Chang, ed., *Chinese Environmental Humanities: Practices of Environing at the Margins* (New York, NY: Springer, 2019); Riccardo Moratto, ed., *Ecocriticism and Chinese Literature: Imagined Landscapes and Real Lived Spaces* (New York, NY: Routledge, 2022).

16. In his Chinese-language monograph *Shengtaipipingyu shengtai sixiang*, mainland Chinese ecocritic Wang Nuo argues that Chinese ecocriticism should transcend the anthropocentric notion and embrace a biocentric view. He suggests that the core of environmental ideas lies in the clash between anthropocentrism and biocentrism. For Wang, Chinese environmental ideas become a prescriptive, ahistorical, and teleological destination. The present book shows that modern Chinese environmentalism should not be reduced to abstract and decontextualized notions. Wang Nuo 王诺, *Shengtaipipingyu shengtai sixiang* 生态批评与生态思想 [Ecological thought and ecocriticism] (Beijing: Renmin chubanshe, 2013).

17. Mark Elvin, *The Retreat of the Elephants: An Environmental History of China* (New Haven, CT: Yale University Press, 2008); Lillian M. Li, *Fighting Famine in North China: State, Market, and Environmental Decline, 1690s–1990s* (Stanford, CA: Stanford University Press, 2007); Judith Shapiro, *Mao's War against Nature: Politics and the Environment in Revolutionary China* (New York, NY: Cambridge University Press, 2001).

18. E. Elena Songster, *Panda Nation: The Construction and Conservation of China's Modern Icon* (Oxford: Oxford University Press, 2018); Sigrid Schmalzer, *Red Revolution, Green Revolution: Scientific Farming in Socialist China* (Chicago: University of Chicago Press, 2016); Miriam Gross, *Farewell to the God of Plague: Chairman Mao's Campaign to Deworm China* (Berkeley: University of California Press, 2016); Joshua Goldstein, *Remains of the Everyday: A Century of Recycling in Beijing* (Berkeley: University of California Press, 2021).

19. Shapiro, *Mao's War*; Frank Dikötter, *Mao's Great Famine: The History of China's Most Devastating Catastrophe, 1958–1962* (New York, NY: Bloomsbury, 2010).

20. For a few examples of how rhetorics function in shaping and understandings Chinese history, see David Strand, *An Unfinished Republic: Leading by Word and Deed in Modern China* (Berkeley: University of California Press, 2011); Elizabeth Perry, *Anyuan: Mining China's Revolutionary Tradition* (Berkeley: University of California Press, 2012); Paul Cohen, *History in Three Keys: The Boxers as Event, Experience, and Myth* (New York, NY: Columbia University Press, 1997).

21. Ole Bruun, *Fengshui in China: Geomantic Divination between State Orthodoxy and Popular Religion* (Honolulu: University of Hawai'i Press, 2003).

22. In late imperial China, starting in the nineteenth century, feng shui was invoked during times of decreasing fuel and resources. County magistrates banned tree cutting to avoid damaging the feng shui of a region. Tristan Brown, *Laws of the Land: Fengshui and the State in Qing Dynasty China* (Princeton, NJ: Princeton University Press, 2023), 225.

23. Meng Fan 梦梵, ed., *Shijing* 诗经 [Book of songs] (Beijing: Tuanjie chubanshe, 1995), 62–63. See also Elvin, *Retreat of the Elephants*.

24. Zhang Lianwei 张连伟, Li Fei 李飞, and Zhou Jingyong 周景勇, eds., *Zhongguo gudai linyewenxian xuandu* 中国古代林业文献选读 [Selected readings from ancient Chinese forestry literature] (Beijing: Zhongguo yanshan chubanshe, 2015), 55; translation adapted from Charles Muller, "Mencius (Selections)," Resources for East Asian Language and Thought, accessed March 12, 2021, http://www.acmuller.net/con-dao/mencius.html. Ancient Chinese philosophers show signs of environmental values. Ian M. Miller argues that the Chinese conception of trees is influenced by different philosophical schools, including Legalism, Confucianism, and Daoism, between the sixth and second centuries BCE. As a result of this contestation of political ideas, state forestry failed to gain traction. Ian M. Miller, "Forestry and the Politics of Sustainability in Early China," *Environmental History* 22, no. 4 (2017): 594–617.

25. Zhang et al., *Zhongguo*, 55.

26. Zhang et al., *Zhongguo*, 197–98.

27. Ian M. Miller, *Fir and Empire: The Transformation of Forests in Early Modern China* (Seattle: University of Washington Press, 2020). Brian Lander offers a material aspect of early China's deforestation and silviculture. See Brian Lander, "Deforestation in Early China: How People Adapted to Wood Scarcity," in *The Cultivated Forest: People and Woodlands in Asian History*, ed. Ian M. Miller et al. (Seattle: University of Washington Press, 2022), 1–19.

28. Meng Zhang, *Timber and Forestry in Qing China: Sustaining the Market* (Seattle: University of Washington Press, 2021), 4–5.

29. Wang Taiyue 王太岳, "Tongshanyin" 铜山吟 [Lament for copper-bearing

hills], in *Qingshiduo* 清诗铎 [Qing bell of poesy], ed. Zhang Yingchang 张应昌 (Beijing: Zhonghuashuju, 1960), 926–28.

30. Jonathan Schlesinger, *A World Trimmed with Fur: Wild Things, Pristine Places, and the Natural Fringes of Qing Rule* (Stanford, CA: Stanford University Press, 2017); David Bello, *Across Forest, Steppe, and Mountain: Environment, Identity, and Empire in Qing China's Borderlands* (New York, NY: Cambridge University Press, 2015).

31. David Bello, "Splintered Habitats: The Fragmentation of Ecotone Northern China's Imperial Woodland Complexes," in *The Cultivated Forest: People and Woodlands in Asian History*, ed. Ian M. Miller et al. (Seattle: University of Washington Press, 2022), 95–119. See also Nicholas K. Menzies, *Forest and Land Management in Imperial China* (New York, NY: St. Martin's Press, 1994); Stanley Richardson, *Forests and Forestry in China: Changing Patterns of Resource Development* (Washington, DC: Island Press, 1990).

32. Robert Pogue Harrison, *Forests: The Shadow of Civilization* (Chicago: University of Chicago Press, 1992).

33. Jeffrey K. Wilson, *The German Forest: Nature, Identity, and the Contestation of a National Symbol, 1871–1914* (Toronto: University of Toronto Press, 2012).

34. John Knott, *Imagining the Forest: Narratives of Michigan and the Upper Midwest* (Ann Arbor: University of Michigan Press, 2012), 8. For another succinct summary of forestry in Western history, see Michael Pollan, *Second Nature: A Gardener's Education* (New York, NY: Grove Press, 1991), 150–75. For writings about trees in the broader Atlantic world between 1660 and 1830, see also Laura Auricchio, Elizabeth Heckendorn Cook, and Giulia Pacini, eds., *Invaluable Trees: Cultures of Nature, 1660–1830* (Oxford: Voltaire Foundation, 2012).

35. Nash, *Wilderness*, 141–60.

36. Eric Rutkow, *American Canopy: Trees, Forests, and the Making of a Nation* (New York, NY: Simon and Schuster, 2012), 254.

37. Neil Maher, *Nature's New Deal: The Civilian Conservation Corps and the Roots of the American Environmental Movement* (New York, NY: Oxford University Press, 2008).

38. For trees and subject identity formation in the US context, see Knott, *Imagining the Forest*; Pollan, *Second Nature*, 150–75; Nash, *Wilderness*.

39. Ling Daoyang 凌道扬, *Ling Daoyang quanji* 凌道扬全集 [Complete works of Ling Daoyang], ed. Liu Zhongguo 刘中国 (Hong Kong: Gongyuan chubanyouxiangongsi, 2009), 23.

40. Ouyang Qin 欧阳勤, Li Zhengfeng 李正峰, and Ouyang Ping 欧阳屏, *Heisedang'an: Teda zainan shijian jishi* 黑色档案：特大灾难事件纪实 [Black archives: documentary of major disasters] (Shenzhen: Haitianchubanshe, 1999), 213.

41. Hao Jingsheng 郝景盛, *Senlin wannenglun* 森林万能论 [The omnipotence of

forests] (Shanghai: Zhengzhong shuju, 1947), 1–9. The book was first written in 1944 but not published until 1947.

42. Haiyan Lee, *Revolution of the Heart: A Genealogy of Love in China, 1900–1950* (Stanford, CA: Stanford University Press, 2005), 15–16; Eugenia Lean, *Public Passions: The Trial of Shi Jianqiao and the Rise of Popular Sympathy in Republican China* (Berkeley: University of California Press, 2007), 4.

43. For a recent study on environmental exercises and China's global ambition, see Annah Lake Zhu, *Rosewood: Endangered Species Conservation and the Rise of Global China* (Cambridge, MA: Harvard University Press, 2022).

44. Bryan Norton, "Environmental Ethics and Weak Anthropocentrism," *Environmental Ethics* 6, no. 2 (1984): 131–48 passim. There is a rise in scholarship considering trees as beings capable of voluntary thinking or possessing unique forms of consciousness. Michael Marder's *Plant-Thinking: A Philosophy of Vegetal Life* is a seminal work in the field of critical plant studies. Tracing Western traditions and continental philosophies, he explores how interacting with plants can transform human thinking, making it more plantlike and less human. Michael Marder, *Plant-Thinking: A Philosophy of Vegetal Life* (New York, NY: Columbia University Press, 2013). See also Peter Wohlleben, *The Hidden Life of Trees: What They Feel, How They Communicate; Discoveries from a Secret World*, trans. Jane Billinghurst (Vancouver: Greystone Books, 2016).

45. Imanuel Noy-Meir, "Desert Ecosystems: Environment and Producers," *Annual Review of Ecology and Systematics* 4, no. 1 (1973): 25–51.

46. The most recent collection is Stephen Brain and Viktor Pál, eds., *Environmentalism under Authoritarian Regimes: Myth, Propaganda, Reality* (New York, NY: Routledge, 2019). Authoritarian governments' power enables them to mobilize the public. See D. J. C. Shearman and J. W. Smith, *The Climate Change Challenge and the Failure of Democracy* (Westport, CT: Praeger, 2007); Bruce Gilley, "Authoritarian Environmentalism and China's Response to Climate Change," *Environmental Politics* 21, no. 2 (2012): 287–307; Scott Moore, "Modernisation, Authoritarianism, and the Environment: The Politics of China's South-North Water Transfer Project," *Environmental Politics* 23, no. 6 (2014): 947–64. See also Heejin Han, "Authoritarian Environmentalism under Democracy: Korea's River Restoration Project," *Environmental Politics* 24, no. 5 (2015): 810–29; Sarah Eaton and Genia Kostka, "Authoritarian Environmentalism Undermined? Local Leaders' Time Horizons and Environmental Policy Implementation in China," *China Quarterly* 218 (June 2014): 359–80; Geoffrey Chen and Charles Lees, "The New, Green, Urbanization in China: Between Authoritarian Environmentalism and Decentralization," *Chinese Political Science Review* 3, no. 2 (2018): 212–31.

47. To name a few recent examples, Yifei Li and Judith Shapiro argue that China adopts environmental practices to implement and reinforce its social control, con-

tributing to what they term *environmental authoritarianism*. Yifei Li and Judith Shapiro, *China Goes Green: Coercive Environmentalism for a Troubled Planet* (New York, NY: Polity Press, 2020), 188. Yanzhong Huang highlights the constraints and flaws in China's approach to its environmental health crisis, such as state-society asymmetry in information and a lack of institutionalized public participation. Yanzhong Huang, *Toxic Politics: China's Environmental Health Crisis and Its Challenge to the Chinese State* (New York, NY: Cambridge University Press, 2020). Iza Ding demonstrates how the Chinese state employs language, symbols, and gestures of good governance to manage public opinions and mitigate crises. Iza Ding, *The Performative State: Public Scrutiny and Environmental Governance in China* (Ithaca, NY: Cornell University Press, 2022).

48. There is increasing interest in shedding light on the rational environmental policies of authoritarian regimes. See Douglas R. Weiner, *Models of Nature: Ecology, Conservation, and Cultural Revolution in Soviet Russia* (Bloomington: Indiana University Press, 2000); Stephen Brain, *Song of the Forest: Russian Forestry and Stalinist Environmentalism, 1905–1953* (Pittsburgh, PA: University of Pittsburgh Press, 2011); Arvid Nelson, *Cold War Ecology: Forests, Farms, and People in the East German Landscape, 1945–1989* (New Haven, CT: Yale University Press, 2005).

49. The understanding of authoritarianism in general is fruitful and pivotal especially after Xi Jinping came to power. To name a few examples, political scientist Minxin Pei delineates the essential attributes of Chinese authoritarianism as the Chinese Communist government's firm grasp on political authority, stringent regulation of media and civil society, and limitations on political liberties. Minxin Pei, *China's Trapped Transition: The Limits of Developmental Autocracy* (Cambridge, MA: Harvard University Press), 2006. On the other hand, Andrew J. Nathan highlights the adaptive capacity of authoritarian China during the leadership transition from Jiang Zemin to Hu Jintao. He notes that the increase in institutional complexity, autonomy, and coherence has bolstered China's ability to navigate evolving challenges effectively, aligning with Samuel P. Huntington's views on political order. Andrew J. Nathan, "Authoritarian Resilience, *Journal of Democracy* 14, no. 1 (2003): 6–17; Samuel P. Huntington, *Political Order in Changing Societies* (New Haven, CT: Yale University Press, 1968), 12. In a more recent context, sociologist Alvin So identifies three defining aspects of Xi Jinping's evolving authoritarianism: the resurgence of the Communist Party, the advancement of a statist approach to the economy, and the renewed emphasis on ideology's role in shaping policy decisions. The present book delves into the nature of Chinese authoritarianism, offering insights from a cultural and historical perspective on its interactions with the environment. It shows that the enduring nature of this governance style can be attributed to its multifaceted approach to environmental issues, marked by a capacity for often imperceptible integration of conflicting notions. Alvin So,

"The Rise of Authoritarianism in China in the Early 21st Century," *International Review of Modern Sociology* 45, no. 1 (2019): 49–70.

50. If Xiamen's chemical plant crisis in 2007 exposes the weakness caused by lack of political participation in authoritarian regimes, the US water crisis in Flint, Michigan, reveals one of the drawbacks of democratic governments. In order to save money, in April 2014, the Flint city government shifted its water source from Lake Huron to the Flint River, which caused lead to leak into the water system. Although both the federal and state governments promised to help in the face of public outrage, the problem persisted until 2019.

51. It should be noted that Civilian Conservation Corps took grassroots opinions into serious consideration, giving rise to a productive mutual interplay between the state and the common people in decision-making. See Maher, *Nature's New Deal*.

52. Per G. Fredriksson and Jim R. Wollscheid, "Democratic Institutions versus Autocratic Regimes: The Case of Environmental Policy," *Public Choice* 130, no. 3 (2007): 381–93; Bo Rothstein and Jan Teorell, "What Is Quality of Government? A Theory of Impartial Government Institutions," *Governance* 21, no. 2 (2008): 165–90; Carlos Wing-Hung Lo et al., "Unpacking the Complexity of Environmental Regulatory Governance in a Globalizing World: A Critical Review for Research Agenda Setting," *Journal of Environmental Policy and Planning* 22, no. 5 (2020): 594–607.

53. Kate Brown, *Plutopia: Nuclear Families, Atomic Cities, and the Great Soviet and American Plutonium Disasters* (Oxford: Oxford University Press, 2013).

54. A 2023 scientific report claims that the lack of local participation jeopardizes the success of this ambitious tree-planting effort. Junjie Zhai et al., "Assessing the Effects of China's Three-North Shelter Forest Program over 40 Years," *Science of the Total Environment* 857 (2023): 9.

55. Zhou Qunfeng 周群峰, "Qingdao 'zhongshushizhang' Zhang Xinqi tuixiuhoubeicha, chuxihuodongshi yipianxusheng" 青岛"种树市长"张新起退休三年后被查，出席活动时市民一片嘘声 [Qingdao "tree-planting mayor" Zhang Xinqi was investigated three years after retiring, and the citizens booed when he attended the event], *Zhongguo xinwenzhoukan* 中国新闻周刊, March 12, 2021, https://www.chinanews.com/gn/2021/03-12/9430456.shtml.

56. The campaign style of tree planting is not an invention of Communist China. During the Republican era, a 1928 article in *News* argued that the tree-planting campaign had not achieved its intended results. This shortfall was attributed to the campaign's inability to popularize forestry knowledge and ensure the protection of trees after planting among the general public. Consequently, the initiative remained an ephemeral movement rather than becoming an everyday practice embraced by the community. Xun Min 勋民, "Zhishu yundong" 植树运动 [Tree-planting movement], *Xinwen bao* 新闻报, March 12, 1928.

57. In 2007, two chemical plants producing para-xylene and terephthalic acid were set to be built in Xiamen, a coastal city in southeastern China. Their construction unsettled around ten thousand residents and caused intense tensions between the government and its citizens. Though Xiamen's local government finally made a concession to the residents, the voices from the grass roots were hardly reported by state-controlled media. See Elizabeth Brunner, *Environmental Activism, Social Media, and Protest in China: Becoming Activists over Wild Public Networks* (New York, NY: Lexington Books, 2019), 55–80.

58. Ah Cheng 阿城, *Shuwang* 树王 [The king of trees], in *Qiwang, Shuwang, Haiziwang* 棋王，树王，孩子王 (Taipei: Dadi Chubanshe, 2007), 71–130.

Chapter 1

1. Gao Chunxian 高春先, "Gucheng zhongxin xiaoxue 3.12 zhishujie huodong baodao" 古城中心小学3.12植树节活动报道 [Report on the Arbor Day activity at Gucheng Central Primary School on March 12], *Nanjingshi Gucheng xiaoxue* 南京市小学网, March 13, 2013, http://www.njsxd.org/news_detail.asp?id=308965.

2. Lydia Liu, *Translingual Practice: Literature, National Culture, and Translated Modernity; China, 1900–1937* (Stanford, CA: Stanford University Press, 1995); Margherita Zanasi, *Saving the Nation: Economic Modernity in Republican China* (Chicago: University of Chicago Press, 2006); Ruth Rogaski, *Hygienic Modernity: The Meanings of Health and Disease in Treaty Port China* (Berkeley: University of California Press, 2004); Kate Merkel-Hess, *The Rural Modern: Reconstructing the Self and State in Republican China* (Chicago: University of Chicago Press, 2016).

3. "Zhishu xueshu" 植树学塾 [Tree-planting school], *Wanguo gongbao* 万国公报, no. 502 (1878): 2–3.

4. "Zhengshi: Xuzhishuxueshu" 政事：续植树学塾 [Political affairs: continuing the tree-planting school], *Wanguo gongbao* 万国公报, no. 504 (1878): 3–4.

5. For Kang Youwei's and Zhang Zhidong's arguments on forestry, see Xiong Datong 熊大桐, *Zhongguo jindai linye shi* 中国近代林业史 [Modern forestry history in China] (Beijing: Zhongguo linye chubanshe, 1994), 79–83.

6. For a local forester, see Pitts's finding on Cui Hu, a forestry school's instructor in Manchuria, who claimed that a national planting day would serve as an essential step in reconstructing Manchurian forestland in 1911. Larissa Pitts, "Unity in the Trees: Arbor Day and Republican China, 1915–1927," *Journal of Modern Chinese History* 13, no. 2 (2019): 301–2.

7. US foresters were inspired by European forestry. See Radkau, *Nature and Power*, 212–20.

8. Han An 韩安, "Shijie geguo guoyou senlin dashi" 世界各国国有森林大势 [The state of state-owned forests in various countries worldwide], *Nonglingongbao* 农林公报 2, no. 3 (1913): 135–43; Han An 韩安, "Shijie geguo guoyou senlin dashi" 世界各

国国有森林大势 [The state of state-owned forests in various countries worldwide], *Nonglingongbao* 农林公报 2, no. 4 (1913): 93–98; Han An 韩安, "Shijie geguo guoyou senlin dashi" 世界各国国有森林大势 [The state of state-owned forests in various countries worldwide], *Nonglingongbao* 农林公报 2, no. 5 (1913): 131–38.

9. *Department of Forestry 1912–1913*, Department of Forestry Catalogue (1900–1914 Yjk81 A2 1900–1914), 11, Manuscripts and Archives, Yale University Library..

10. *Department of Forestry*, 12–13.

11. Michael B. Smith, "The Value of a Tree: Public Debates of John Muir and Gifford Pinchot," *Historian* 60, no. 4 (1998): 757–78; Char Miller, *Gifford Pinchot and the Making of Modern Environmentalism* (Washington, DC: Island Press, 2013).

12. Student Record of School of Forestry, 1912–14, Manuscripts and Archives, Yale University Library.

13. Thanks to US missionaries, Arbor Day gained momentum in Japan and, later, in Japan's colony Korea, becoming a means of nurturing engaged and responsible citizens and assimilating colonial subjects. American preacher and diplomatic envoy Birdsey Grant Northrop advocated combining rural improvement and youth education with reforestation in Japan. Northrop influenced vice minister of education Makino Nobuaki, who believed that forests were beneficial for the economy as well as for nurturing love of one's nation. Nobuaki later promoted student-led tree-planting activities. In April 1927, Japan held its Arbor Day ceremony in Tokyo, in which more than eight thousand children participated. During Arbor Day, the Japanese promoted the idea that trees are the lungs of the city and could beautify it. In colonial Korea, the Japanese Empire advanced "forest-love thought" among its subjects, resulting in national Arbor Days. These tree-planting initiatives drove home the message of self-sacrifice and state-sanctioned consumption among the Koreans. See "Jushokusai ni atsumatta utsukushii hassen no hitobito, ame no fuzei mo hitoiori fusawashiku, kyou Hibiya no seikan" 樹植祭に集まった美しい８０００の人々：雨の風情も一入ふさわしくけふ日比谷の盛観 [The beautiful 8,000 people gathered at Arbor Day: the rain added a fitting atmosphere to today's grand spectacle at Hibiya], *Asahi shinbun* 朝日新聞, April 4, 1928. See also Fedman, *Seeds of Control*, 176.

14. Joseph Bailie papers, University of Nanking [Nanjing], box 204, folder 3473, Yale Divinity School Archive.

15. For a study of Bailie's Colonization Association, see Randall E. Stross, *The Stubborn Earth: American Agriculturalists on Chinese Soil, 1898–1937* (Berkeley: University of California Press, 1986), 66–92.

16. Bailie strove to convert naked Chinese mountains into a valuable asset in the form of forests, which could further stabilize Chinese society in age of social chaos. See Joseph Bailie, "Famine and Afforestation in China," *International Review of Mission* 6, no. 3 (1917): 450–56.

17. Joseph Bailie, "Magna Charta of China's Forestry Work," *American Forestry* 22 (May 1916): 271–72.

18. Bailie, "Magna Charta," 272.

19. For American foreign policy in China during the First World War, see Robert E. Hannigan, *The Great War and American Foreign Policy, 1914–24* (Philadelphia: University of Pennsylvania Press, 2017), 79–92.

20. The appropriation of Western thinking on trees in China is comparable to the introduction of new ideas concerning animals in Japan. Brent Walker describes how the Meiji government hired the American Edwin Dun to help eradicate Japan's wolves when the Western-influenced Japanese realized that wolves spread the rabies virus. Both cases show how the Western scientific conception of the natural world changed individuals' mindsets and environmental policies in East Asia. See Brett Walker, *The Lost Wolves of Japan* (Seattle: University of Washington Press, 2005).

21. Ling Daoyang, *Ling Daoyang*, 3.

22. "Cheng dazongtongwen: Niding qingming wei zhishujie qingyi shenling xuanshi" 呈大总统文: 拟定清明为植树节请以申令宣示 [To the president: proposing Qingming as Arbor Day, please declare by presidential order], *Nongshang gongbao* 农商公报 2, no. 1 (1915): 7.

23. An Shinong 安事农, *Zhongguo senlinfa* 中国森林法 [Chinese forestry law] (Shanghai: Huatong shuju, 1933); Wu Jinzan 吴金赞, *Zhonghua minguo linye fazhishi* 中华民国林业法制史 [Forestry legal system of the Republic of China] (Taipei: Zhengzhong shuju, 1991).

24. "Xishan zhishujie zhi shenghui" 西山植树节之盛会 [The grand celebration of Arbor Day at Xishan], *Yishibao (Tianjinban)* 益世报 (天津版), April 7, 1916.

25. "Dazongtong junjian guowuyuan neiwu jiaoyunongshang gezongzhang junjian bennian zhishujie yejing dianling daoxian gexuexiao yili juxing zhishu jinian" 大总统钧鉴国务院内务教育农商各总长钧鉴: 本年植树节业经电令道县各学校一例举行植树纪念 [To the president's august attention, to the state council's ministers of internal affairs, education, agriculture, and commerce: this year's Arbor Day activities have been ordered by telegram to be uniformly conducted as a commemorative event in schools at all levels across the prefectures and counties], *Zhengfu gongbao* 政府公报 450 (1917): 21.

26. Ling Daoyang, *Ling Daoyang*, 9.

27. Liu Zongyuan claims that to ensure the growth of a specific tree, one needs to respect the living patterns of the plants by leaving them alone. Liu Zongyuan 柳宗元, "Zhongshu Guo Tuotuo zhuan" 种树郭橐驼传 [The legend of Guo Tuotuo planting trees], in *Zhongguo gudai linye wenxian xuandu* 中国古代林业文献选读 [Selected readings from ancient Chinese forestry literature], ed. Zhang Lianwei 张连伟, Li Fei 李飞, and Zhou Jingyong 周景勇 (Beijing: Beijing yanshan chubanshe, 2015), 196–99.

28. Yuan Zhen 元稹, "Duixiudiqingzhongshupan" 对修堤请种树判 [Judgment on the request to plant trees for dike repair], in *Zhongguo gudai linye wenxian xuandu* 中国古代林业文献选读 [Selected readings from ancient Chinese forestry literature], ed. Zhang Lianwei 张连伟, Li Fei 李飞, and Zhou Jingyong 周景勇 (Beijing: Beijing yanshan chubanshe, 2015), 193–94.

29. Parks were widely promoted in the United States at the beginning of the twentieth century, reflecting the belief that they lead to better citizens. See Matthew Klingle, *Emerald City: An Environmental History of Seattle* (New Haven, CT: Yale University Press, 2007), 119–53.

30. Ling Daoyang 凌道扬, *Senlinxue dayi* 森林学大意 [Main ideas of forestry] (Shanghai: Shangwu yinshuguan, 1916), 40. For the English version, see Dau-yang Lin [Ling Daoyang 凌道扬], *Chapters on China and Forestry* (Shanghai: Commercial Press, 1916).

31. Ling Daoyang 凌道扬, *Senlin yaolan* 森林要览 [Forest outline] (Shanghai: Shangwu yinshuguan, 1918).

32. Chi Xiang 池翔, "Chongsubianjiang: Yalujiang youan de linmu caifa senlinjiaosheyubianjiangzhixu" 重塑边疆:鸭绿江右岸的林木采伐森林交涉与边疆秩序 [Reshaping the frontier: timber harvesting, forest negotiations, and frontier order on the right bank of the Yalu River], *Chongqingdaxuexuebao* 重庆大学学报5 (2020): 265–78; Wang Xiliang 王希亮, "Jindai zhongguodongbei senlin de zhiminkaifa yu shengcunkongjian bianqian" 近代中国东北森林的殖民开发与生态空间变迁 [Colonial development and ecological spatial changes in forests in northeastern China in modern times], *Lishiyanjiu* 历史研究1 (2017): 85–103; Zheng Yu 郑宇, *Jindai dongbei senlin ziyuan chanyehua yanjiu (1878–1931 nian)* 近代东北森林资源产业化研究 (1878–1931年) [Modern northeastern forestry resource industrialization study (1878–1931)] (Shanghai: Shanghai shehuikexue chubanshe, 2020). See also Ruth Rogaski, *Knowing Manchuria: Environments, the Senses, and Natural Knowledge on an Asian Borderland* (Chicago: University of Chicago Press, 2022).

33. Yasutomi Ayumu 安冨步 and Fukao Yōko 深尾葉子, eds., *"Manshū" no seiritsu: Shinrin no shōjin to kindai kūkan no keisei* "満洲"の成立: 森林の消尽と近代空間の形成 [The formation of "Manchuria": forest depletion and the creation of modern spaces] (Nagoya-shi: Nagoya daigaku shuppankai, 2009), 19–60.

34. "Zhonghuasenlinhui zhi guohui geyiyuanshu" 中华森林会致国会各议员书 [Letter from the Chinese Forestry Association to the members of the national congress], *Shenbao* 申报, April 8, 1917.

35. Ling Daoyang, *Ling Daoyang*, 1.

36. For a vivid account of the three principles of the people, see Marie-Claire Bergère, *Sun Yat-sen*, trans. Janet Lloyd (Stanford, CA: Stanford University Press, 1998), 352–94. See also Audrey Wells, *The Political Thought of Sun Yat-sen: Development and Impact* (New York, NY: Palgrave, 2001).

37. Sun Zhongshan [Sun Yat-sen] 孙中山, "Sanminzhuyi minshengzhuyi" 三民主义民生主义 [The three principles of the people, the principle of people's livelihood], in *Sunzhongshan quanji* 孙中山全集 [Complete works of Sun Yat-sen], vol. 9, ed. Guangdongsheng shehuikexueyuan 广东省社会科学院, Zhongguo shekeyuan 中国社科院, and Zhongshan daxue 中山大学 (Beijing: Zhonghuashuju, 1986), 408.

38. Merkel-Hess, *Rural Modern*. According to Merkel-Hess, intellectuals in the Republican era advocated a long, slow process of self-transformation, mass mobilization, and bottom-up state creation in rural China.

39. Zhou Hui'an 周晦盦, *Zhengfu ziran* 征服自然 [Conquering nature] (Beiping: Xiehua chubanshe, 1932), 29.

40. Zhou Hui'an 周晦盦, *Zhishu jie* 植树节 [Arbor Day] (Beiping: Xiehua chubanshe, 1932).

41. Sun Zhongshan [Sun Yat-sen] 孙中山, "Jianguo fanglüe" 建国方略 [Plan for building the nation], in *Sunzhongshan quanji* 孙中山全集 [Complete works of Sun Yat-sen], vol. 6, ed. Guangdongsheng shehuikexueyuan 广东省社会科学院, Zhongguo shekeyuan 中国社科院, and Zhongshan daxue 中山大学 (Beijing: Zhonghuashuju, 1985), 389.

42. Sun, "Jianguo fanglüe," 390–94.

43. Shellen Wu, *Empires of Coal: Fueling China's Entry into the Modern World Order, 1860–1920* (Stanford, CA: Stanford University Press, 2015).

44. Sun Zhongshan [Sun Yat-sen] 孙中山, "Zhongguo zhi tielu jihua yu minshengzhuyi" 中国之铁路计划与民生主义 [China's railway plan and principles of people's livelihood], in *Sunzhongshan quanji* 孙中山全集 [Complete works of Sun Yat-sen], vol. 2, ed. Guangdongsheng shehuikexueyuan 广东省社会科学院, Zhongguo shekeyuan 中国社科院, and Zhongshan daxue 中山大学 (Beijing: Zhonghuashuju, 1982), 493.

45. Sun Zhongshan [Sun Yat-sen] 孙中山, "Zai shanghai Zhongguo shehuidangde yanshuo" 在上海中国社会党的演说 [Speech at the Chinese Socialist Party in Shanghai], in *Sunzhongshan quanji* 孙中山全集 [Complete works of Sun Yat-sen], vol. 2, ed. Guangdongsheng shehuikexueyuan 广东省社会科学院, Zhongguo shekeyuan 中国社科院, and Zhongshan daxue 中山大学 (Beijing: Zhonghuashuju, 1982), 518. Sun Yat-sen's expressions are within the contours of his understanding of the socialist movement, which he supported, albeit with reservations. He argued that conflicts arose in European countries because of unequal wealth distribution; China should try to avoid such class conflicts. One way of achieving this goal was to involve the workers rather than the tycoons in forestry development. Sun's amenability to political participation in forestry had significant implications: workers were no longer passive objects to be manipulated; rather, they were agents of nation building, including forestry management.

46. Sun Zhongshan [Sun Yat-sen] 孙中山, "Zhongguo guomindang xuanyan" 中

国国民党宣言 [Chinese Nationalist Party declaration], in *Sunzhongshan quanji* 孙中山全集 [Complete works of Sun Yat-sen], vol. 7, ed. Guangdongsheng shehuikexueyuan 广东省社会科学院, Zhongguo shekeyuan 中国社科院, and Zhongshan daxue 中山大学 (Beijing: Zhonghuashuju, 1985), 4.

47. Ian Tyrrell, *Crisis of the Wasteful Nation: Empire and Conservation in Theodore Roosevelt's America* (Chicago: University of Chicago Press, 2015).

48. For two empirical studies of China's deforestation and social dynamics from the 1910s to the 1930s, see Kenneth Pomeranz, *The Making of a Hinterland: State, Society, and Economy in Inland North China, 1853–1937* (Berkeley: University of California Press, 1993), 145–46; Peter C. Perdue, *Exhausting the Earth: State and Peasant in Hunan, 1500–1850* (Cambridge, MA: Harvard University Asia Center, 1987), 234–35.

49. Wang Pixian 王丕显, "Duiyu zhishujie yingxing zhengdun zhi yijian" 对于植树节应行整顿之意见 [Suggestions to rectify Arbor Day], *Jiangsu shiye yuezhi* 江苏实业月志, no. 4 (1926): 118–23.

50. Huang Taiming 黄泰明, "Lun Qingming zhishu shifou xiangyi" 论清明植树是否相宜 [On whether it is appropriate to plant trees during the Qingming festival], *Xuesheng wenyi congkan* 学生文艺丛刊 3, no. 6 (1926): 49–50.

51. Bao Shen 葆琛, "Zhishujie de laili" 植树节的来历 [The origin of Arbor Day], *Nongmin* 农民2, no. 4 (1926): 4–6.

52. Zhang Ziping 张资平, *Zhishujie* 植树节 [Arbor Day] (Shanghai: Xinyuzhou shudian, 1928), 22.

53. Xun, "Zhishu yundong."

54. "Gaiding chunfen wei zhishujie" 改定春分为植树节 [Change Arbor Day to the spring equinox], *Guoli disanzhongshandaxue jiaoyu zhoukan* 国立第三中山大学教育周刊 21 (1928): 5.

55. "Zhishu yi jinian zhongshanxiansheng" 植树以纪念中山先生 [Plant a tree in memory of Sun Yat-sen], *Shibao* 时报, March 12, 1928.

56. Wu Shuchen 吴庶晨, "Zhishushi yu zhishujie" 植树式与植树节 [Tree-planting ceremony and Arbor Day], *Xinwen bao* 新闻报, March 12, 1929.

57. "Gebu jinian sunzongli, tongri juxing zhishujie" 各埠纪念孙总理, 同日举行植树节 [Various towns commemorate Premier Sun, hold Arbor Day on the same day], *Shishixinbao (Shanghai)* 时事新报(上海), March 13, 1929. In the northeastern part of China, Arbor Day was not observed on March 12 because it was too cold. "Zhishujie dianlinggaiqi" 植树节电令改期 [Arbor Day telegraphic order for rescheduling], *Shengjingshibao* 盛京时报, March 22, 1929.

58. Yang Fu 仰夫, "Zhishu yu chuxu" 植树与储蓄 [Planting and saving money], *Xinwen bao* 新闻报, March 12, 1928.

59. Huang Yingdai 黄影呆, "Zhishujie ganyan" 植树节感言 [Arbor Day speech], *Xinwen bao* 新闻报, March 14, 1933.

60. Zhao Jiting 赵骥亭, "Lun zhishujie zhi youming wushi" 论植树节之有名无实 [On Arbor Day being in name only, not in reality], *Yingqiu* 嘤求, no. 4 (1928): 6–8.

61. "Lingmu qian zhishu dianli" 陵墓前植树典礼 [Tree-planting ceremony in front of the mausoleum], *Minguo ribao* 民国日报, March 13, 1929.

62. Henrietta Harrison, *The Making of the Republican Citizen: Political Ceremonies and Symbols in China 1911–1929* (New York, NY: Oxford University Press, 2000), 2.

63. George Perkins Marsh, *Man and Nature: or, Physical Geography as Modified by Human Action* (1864; repr., Seattle: University of Washington Press, 2003). For study on global intellectual transfer, see William Kirby, *Empires of Ideas: Creating the Modern University from Germany to America to China* (Cambridge, MA: Harvard University Press, 2022).

64. Marsh, *Man and Nature*, 357.

65. For a history of Arbor Day in the United States, see Cohen, *Planting Nature*, 26–47.

66. Nathaniel Hillyer Egleston, *Arbor Day: Its History and Observance*, Report (US Department of Agriculture) 56 (Washington, DC: US Government Printing Office, 1896), 25. See also Eric Klinek, "Arbor Day: Educating Environmental Citizens," *Pennsylvania Legacies* 10, no. 1 (2010): 3–5.

67. H. Harrison, *Republican Citizen*, 3.

68. "Zongli shishi wuzhounian zhishu jinian: Zhishu qingxing" 总理逝世五周年植树纪念: 植树情形 [Tree-planting commemoration of the fifth anniversary of the death of the prime minister: tree-planting situation], *Nanxun tielu yuebao* 南浔铁路月报 8, no. 3 (1930): 1. For an excellent Chinese-language discussion of Arbor Day's nationwide symbolic events, see Chen Yunqian's powerful case study of Anhui Province. Chen Yunqian 陈蕴茜, "Zhishujie yu sunzhongshan chongbai" 植树节与孙中山崇拜 [Arbor Day and the worship of Sun Yat-sen], *Nanjing daxue xuebao* 南京大学学报 13, no. 5 (May 2006): 79–90.

69. Zhong Ming 中铭, "Zhenwang jiangshi zhuidao dahui zhishudianli" 阵亡将士追悼大会植树典礼 [Tree-planting ceremony at memorial meeting for fallen soldiers], *Shishi yuebao* 时事月报 4, no. 4 (1931): 13.

70. Empirical evidence illustrates that economic investment in forestry was meager after Sun's death. During the Sino-Japanese War, expenditure on forestry only constituted a small component of agriculture and forestry budgets, ranging from one-sixth at the most and one-fortieth at the least. See Xiong et al., *Zhongguo*, 160.

71. Hu Jie'an 胡洁庵, "Zhishu ji" 植树记 [Documenting tree planting], *Anhui jiaoyu* 安徽教育 2, no. 11 (1931): 64–65.

72. Hu Jie'an, "Zhishuji," 64.

73. Pitts, "Unity in the Trees," 296–318.

74. Jiang Jieshi [Chiang Kai-shek] 蒋介石, "Xin shenghuo yundong zhi yaoyi"

新生活运动之要义 [Key ideas of the New Life movement], in *Xin shenghuo yundong* 新生活运动 [The New Life movement], ed. Ye Chucang 叶楚伧 (Nanjing: Zhongzheng shuju, 1934), 8–23.

75. "Zhishu" 植树 [Tree planting], *Guowen zhoubao* 国闻周报11, no. 12 (1934): 1. Chen Guofu, Chiang's intimate political ally, explained that January, February, August, and September are good times to encourage youngsters to study, March is for planting trees, April is for silk production, May is for national awareness, June is for fighting against drugs, July should be devoted to cooperation, August is for artistic movements, November should focus on road construction, and December is to be utilized for public infrastructure construction. See Chi Xinhong 池昕鸿, ed., *Chen Guofu quanzhuan* 陈果夫全传 [Complete works of Chen Guofu] (Yanji: Yanbian renmin chubanshe, 2003).

76. Frank Uekötter, *The Green and the Brown: A History of Conservation in Nazi Germany* (New York, NY: Cambridge University Press, 2006), 25.

77. Chiang Kai-shek was heavily influenced by Germany. See William Kirby, *Germany and Republican China* (Stanford, CA: Stanford University Press, 1984).

78. Maggie Clinton, *Revolutionary Nativism: Fascism and Culture in China, 1925–1937* (Durham, NC: Duke University Press, 2017), 158.

79. *Jiang Jieshi zhishu* 蒋介石植树 [Chiang Kai-shek plants trees], March 12, 1929, photograph, Minguotupian ziyuanku 民国图片资源库, Zhenjiang, Jiangsu Province.

80. Zhu Wenshu 朱文叔 and Lü Boyou 吕伯攸, eds., *Xiaoxueguoyukeben: Chuji* 小学国语课本: 初级 [Primary school Chinese textbook: elementary level] (Shanghai: Zhonghua shuju, 1936), 18–19.

81. Chen Bochui 陈伯吹, Lü Jinlu 吕金录, and Zhuang Yu 庄俞, eds., *Fuxing guoyu keben* 复兴国语课本 [Reviving the Chinese-language textbook], vol. 5 (Shanghai: Shangwu yinshuguan, 1935), 23–24.

82. Emotion—in particular, sympathy—plays an important role in molding the Chinese mind and cultivating a participatory politics. Here, the textbook is using trees to foster sympathetic sentiments among children. For discussions of sympathy and emotion in the first half of the twentieth century, see Lee, *Revolution of the Heart*; Lean, *Public Passions*.

83. The idea of cultivating a tree is used to illustrate the demanding needs and tasks involved in educating a person in ancient China. The seventh-century BCE Chinese philosopher and statesman Guan Zhong claims, "It takes ten years to nurture a tree, but a hundred years to train a man." Guan Zhong 管仲, *Guanzi* 管子 [Master Guan], ed. Li Yuanyan 李元燕 and Li Wenjuan 李文娟 (Guangzhou: Guangzhou chubanshe, 2001), 26. Here, in the Republican era, the act of tree planting and its impact on cultivating a person's character is not merely metaphorical but also presented as mutually constructive.

84. Ma Ganlin 马甘霖, "Zhishujie" 植树节 [Arbor Day], *Ertong shijie (Shanghai 1922)* 儿童世界 (上海1922) 34, no. 5 (1935): 68–71.

85. Guofu shishi shiqizhounian shengshi longzhong juxing jinian wanzhong qianshu tuixing xinguominyundong tichang zhishulühuawuhan" 国父逝世十七周年 省市隆重举行纪念 万众签署推行新国民运动 提倡植树绿化武汉 [Seventeenth anniversary of Sun-Yet-sen's death: provincial and city memorial held magnificently, tens of thousands sign to promote New Citizen Movement advocating afforestation and greening Wuhan," *Dachubao* 大楚报, March 12, 1942; "Shoudu zhishuzaolin yundonghui" 首都造林运动会 [The Tree-planting Movement in the Capital], *Minguoribao nanjingxinbaoban* 民国日报南京新报版, February 27, 1942.

86. The first integration of both Arbor Day and the spiritual mobilization movement happened in 1940. The drama of the Arbor Day celebrations in both Chongqing and Nanjing continued throughout the Second World War. "Zhishujiepeidujuxing zhishudianli" 植树节陪都举行植树典礼 [The capital holds a tree-planting ceremony on Arbor Day], *Xijing shibao* 西京时报, March 13, 1942; "Zaolinzhishu dianli zhongshuyaoren duoqinwang canjia shenbuzhang juchu sanzhongyiyi" 造林植树典礼中枢要人多亲往参加沈部长举出三种意义 [Afforestation and tree-planting ceremony: many important figures from the central government attended in person; Minister Shen elaborates on three significances], *Zhongyang ribao (Chongqing)* 中央日报(重庆), March 13, 1942.

87. Reviewing Sun's legacy, Marie-Claire Bergère shows that his stories are adapted to the needs of political authorities. See Bergère, *Sun Yat-sen*, 408–19.

88. "Jingxuesheng canbai guofu tongxiang" 京学生参拜国父铜像 [Students in Nanjing pay homage to the bronze statue of the founding father], *Shenbao* 申报, March 13, 1943; "Han juxing zhishu dianli" 汉举行植树典礼 [Hankou holds tree-planting ceremony], *Shenbao* 申报, March 13, 1943.

89. Hao, *Senlin*, 1.

90. Hao Jingsheng 郝景盛, "Zhishujie shuguan" 植树节书感 [Reflections on Arbor Day], *Dagongbao* 大公报, March 12, 1948.

91. "Minguo sanshiqinian zhishujie teji: Nonglinbu zuobuzhangshunsheng zhishujie zhuci" 民国三十七年植树节特辑: 农林部左部长舜生植树节祝词 [The Republic of China's thirty-seventh year Arbor Day special: Arbor Day message by Minister Zuo Shunsheng of the Ministry of Agriculture and Forestry], *Linchantongxun* 林产通讯 2, no. 6 (1948): 2.

92. "Youganyu zhishujie" 有感于植树节 [Reflections on Arbor Day], *Shishi xinbao wankan* 时事新报晚刊, March 14, 1948.

93. "Yuesheng fangshui yu zaolin" 粤省防水与造林 [Flood prevention and afforestation in Guangdong Province], *Shenbao* 申报, May 6, 1949.

94. "Shouming zuge de He Yingqin jiangjun" 受命组阁的何应钦将军 [General He Yingqin appointed to form a cabinet], *Xinwen tiandi* 新闻天地, no. 62 (1949): 1.

95. Jiang Xin 江心 et al., "Shaan Gan Ning bianqu linyefazhanshiyanjiu 1937–1950" 陕甘宁边区林业发展史研究1937–1950 [Research on the history of forestry de-

velopment in the Shaanxi-Gansu-Ningxia border region 1937–1950], *Beijing linye daxue xuebao* 北京林业大学学报11, no. 1 (2012): 1–24. See also Yang Shaozhang 杨绍章 and Xin Yejiang 辛业江, *Zhongguo linye jiaoyushi* 中国林业教育史 [History of forestry education in China] (Beijing: Zhongguolinye chubanshe, 1988), 46–49. Yan Xishan controlled Shanxi Province and joined Chiang Kai-shek's Nationalist government in 1927. Nominally, Shanxi was part of the Nationalist territory.

96. From the 1910s to the 1930s, the Beiyang and, later, the Nationalist government set up a number of schools to promote forestry. In 1914, Beiping Agricultural School was established; in 1927, the Central Agricultural School was founded in Nanjing. Shen Guofang 沈国舫 and Liu Jia 刘佳, "Zhongguo gaodeng linyejiaoyu de jiegou jiqi tiaozheng" 中国高等林业教育的结构及其调整 [The structure and adjustment of higher forestry education in China], in *Nongyejiaoyu dexianzhuang he zhanwang guoji taolunhui lunwenji* 农业教育的现状和展望国际讨论会论文集 [Proceedings of the International Symposium on the Current Situation and Prospects of Agricultural Education] (Beijing: Beijing nongyedaxue chubanshe, 1990), 186–93.

97. This global transmission of knowledge is by no means confined to forestry. Sigrid Schmalzer shows that a significant portion of Chinese agricultural knowledge derived from the United States. Schmalzer, *Red Revolution*, 47–72.

98. The Jiangxi Soviet government was weak and vulnerable owing to lack of food, materiel, and financial support and was under imminent threat from the Nationalist government. Mao, however, still advocated tree planting and ordered people not to impede the growth of seedlings. Trees provided fuel for heat as well as preventing flood and drought, which could lead to famine and social instability. "Linshi zhongyangzhengfu wengao: Renmin weiyuanhui duiyu zhishuyundongde jueyian" 临时中央政府文告：人民委员会对于植树运动的决议案 [Provisional central government announcement: resolution of the People's Committee on the Tree-Planting Movement], *Hongse Zhonghua* 红色中华, no. 15 (1932): 6.

99. "Menglie kaizhan guangda zhishu yundong: Wei tigao nongyeshengchan er douzheng, zenyang qu jinxingzhishu yundong" 猛烈开展广大植树运动：为提高农业生产而斗争，怎样去进行植树运动 [Vigorously launch a wide-scale tree-planting campaign: strive to improve agricultural production, how to conduct a tree-planting movement], *Hongse Zhonghua* 红色中华, no. 145 (1934): 2.

100. "Zengzhishumu baohushanlin" 增植树木保护山林 [Plant more trees and protect mountains and forests], *Hongse Zhonghua* 红色中华, no. 150 (1934): 3.

101. "Huichang fandongfenzi fanghuoshaoshan" 会昌反动分子放火烧山 [Huichang reactionaries set fire to the mountain], *Hongse Zhonghua* 红色中华, no. 177 (1934): 2.

102. Jiang Xin 江心, "Zhishu he hulin" 植树和护林 [Tree planting and forest protection], *Jiefang ribao* 解放日报, March 25, 1942.

103. Huang Zhenglin 黄正林, *Shaan Gan Ning bianqu shehuijingjishi (1937–1945)* 陕甘宁边区社会经济史 (1937–1945) [Social and economic history of the Shaanxi-Gansu-Ningxia border region (1937–1945)] (Beijing: Renmin chubanshe, 2006), 28.

104. Huang Zhenglin, *Shaan Gan Ning*, 31–32; Shaan Gan Ning bianqu caizhengjingjishi bianxiezu 陕甘宁边区财政经济史编写组, ed., *Kangrizhanzhengshiqi Shaan Gan Ning bianqu caizhengjingji shiliaozhaibian* 抗日战争时期陕甘宁边区财政经济史料摘编 [Excerpts from financial and economic historical materials of the Shaanxi-Gansu-Ningxia border region during the Anti-Japanese War], vol. 2 (Wuhan: Changjiangwenyi chubanshe, 2016), 116–52.

105. Huang Lide 黄立德 and Zeng Aidi 曾艾狄, "Zhishu de fangfa" 植树的方法 [Methods of planting trees], *Jiefang ribao* 解放日报, April 5, 1943.

106. Jiang Xin 江心, "Senlin yousha yongchu" 森林有啥用处 [What the use of forests is], *Jiefang ribao* 解放日报, September 24, 1944.

107. Jiang Xin 江心, "Senlin de jianjie zuoyong" 森林的间接作用 [Indirect benefits of forests], *Jiefang ribao* 解放日报, September 26, 1944.

108. Jiang Xin 江心, "Yan'anshi zhishu diaocha" 延安市植树调查 [Tree-planting survey of Yan'an], *Jiefang ribao* 解放日报, March 2, 1945.

109. Although technical guidelines on tree planting were published in Nationalist territory during the 1940s, the mainstream media, such as the Nationalist mouthpiece newspaper *Central Daily News* (*Zhongyang ribao*), continued to focus more on the ceremonial aspect of Arbor Day rather than the survival of the trees. A 1946 article in *Central Daily News* entitled "Reflections on Tree Planting" ("Zhishu youguan") suggests that "trees are planted every year, but there are no trees to be seen every year." The author argues that tree planting has become merely a slogan. "Zhishu youguan" 植树有感 [Reflections on tree planting], *Zhongyang ribao* 中央日报, March 14, 1946.

110. "Jinri qingming bianqugedi qunzhong zhunbei daliangzhishu" 今日清明，边区各地群众准备大量植树 [Today is the Qingming festival, and people from all over the border area are preparing to plant a large number of trees], *Jiefang ribao* 解放日报, April 5, 1946.

111. *Guoyu keben* 国语课本 [Mandarin textbook], vol. 3 (n.p.: Huabeixinhuashudian, 1948), 13–15.

112. "Zhishujie jinianhui" 植树节纪念会 [Arbor Day commemoration party], *Lianhebao* 联合报, March 12, 1952.

113. Guan Bingzong 关秉宗, "Zhishu huodong fansi: Que yongxu linye kanshu kandao feizhouqu" 植树活动省思：缺永续林业砍树砍到非洲去 [Reflections on tree-planting activities: lacking sustainable forestry, cutting down trees extends to Africa], *Lianhebao* 联合报, March 14, 2016.

114. Chen Xiaoyuan 陈晓原, "Xinzhongguo de senlin shiye" 新中国的森林事业 [The forestry industry in new China], *Renmin ribao* 人民日报, September 15, 1950.

115. Wu Jing 吴晶 and Yang Yijun 杨依军, "Xi Jinping zai canjia shoudu yiwuzhishu huodong shi qiangdiao: Laogu shuli lüshuiqingshan jiushi jinshanyinshan linian dazao qingshanchangzai lüshuichangliu kongqichangxin meilizhongguo" 习近平在参加首都义务植树活动时强调：牢固树立绿水青山就是金山银山理念打造青山常在绿水长流空气常新美丽中国 [During the capital's voluntary tree-planting activity, Xi Jinping emphasizes that we must firmly establish the concept that lucid waters and lush mountains are invaluable assets and strive to create a beautiful China with evergreen mountains, perennially flowing waters, and ever-fresh air], *Xinhuanet* 新华网, April 3, 2020, http://www.xinhuanet.com/politics/leaders/2020-04/03/c_1125811438.htm.

Chapter 2

1. Liu Wenhui 刘文辉, "Dayuejin zhongde linye jianshe linyebu buzhang liuwenhui de fayan" 大跃进中的林业建设林业部部长刘文辉的发言 [Forestry construction during the Great Leap Forward: speeches by forestry minister Liu Wenhui], *Renmin ribao* 人民日报, April 6, 1960. Liu cites French forester Jean Messines's article in *Unasylva*, a journal published by the Food and Agriculture Organization of the United Nations. See Jean Messines, "Forest Rehabilitation and Soil Conservation in China," *Unasylva* 12, no. 3 (1958): 102–20.

2. For some representative works of this destructive depiction of Mao-era environmental policy, see Shapiro, *Mao's War*; and Dikötter, *Mao's Great Famine*, 174. For two earlier studies of Chinese forestry, see Menzies, *Forest*; Richardson, *Forests*. There is a growing trend toward a revisionist perspective on Maoist environmental rhetoric. See Gross, *Farewell*; Schmalzer, *Red Revolution*; Songster, *Panda Nation*; Christopher Reed, "Sino-silviculture: State-Sponsored Green Forestry Initiatives in Mao's China," in *Environmentalism under Authoritarian Regimes: Myth, Propaganda, and Reality*, ed. Stephen Brain and Viktor Pal (New York, NY: Routledge, 2018), 180–90.

3. For a brief study of German forestry and its global impact, see James C. Scott, *Seeing like a State: How Certain Schemes to Improve the Human Condition Have Failed* (New Haven, CT: Yale University Press, 1998), 11–52.

4. Franz Heske, *German Forestry* (New Haven, CT: Yale University Press, 1938), 213–14.

5. Liang Xi 梁希, "Minshengwenti yu senlin" 民生问题与森林 [The problems of livelihood and the forest], in *Liang Xi wenji* 梁希文集 [Collected works of Liang Xi] (Beijing: Zhongguo linye chubanshe, 1983), 7.

6. Liang Xi 梁希, "*Zhongguo linye* fakanci" 《中国林业》发刊词 [Foreword to *Chinese forestry*], in *Liang Xi wenji* 梁希文集 [Collected works of Liang Xi] (Beijing: Zhongguo linye chubanshe, 1983), 207.

7. Xunzi (ca. 310–235 BCE) is a good example of early Confucian environmental

thought. He called people to regulate their desires and be prudent in their behaviors. On Confucian thought, see Philip J. Ivanhoe, "Early Confucianism and Environmental Ethics," in *Confucianism and Ecology: The Interrelation of Heaven, Earth, and Humans,* ed. Mary Tucker and John Berthrong (Cambridge, MA: Harvard University Press, 1998), 59–76. On the five phases, see Robert P. Weller and Peter K. Bol, "From Heaven-and-Earth to Nature: Chinese Concepts of the Environment and Their Influence on Policy Implementation," in *Confucianism and Ecology: The Interrelation of Heaven, Earth, and Humans,* ed. Mary Tucker and John Berthrong (Cambridge, MA: Harvard University Press, 1998), 313–42.

8. Liang, "*Zhongguo linye* fakanci," 208.

9. Liang, "*Zhongguo linye* fakanci," 208.

10. Ling Zhang, *The River, the Plain, and the State: An Environmental Drama in Northern Song China, 1048–1128* (New York, NY: Cambridge University Press, 2016); Micah S. Muscolino, *The Ecology of War in China: Henan Province, the Yellow River, and Beyond, 1938–1950* (New York, NY: Cambridge University Press, 2014).

11. Ruth Mostern, *The Yellow River: A Natural and Unnatural History* (New Haven, CT: Yale University Press, 2021).

12. David Pietz, *The Yellow River: The Problem of Water in Modern China* (Cambridge, MA: Harvard University Press, 2015), 70–129.

13. Wang Xiqun 王希群 et al., eds., *Zhongguo linye shiye de xianqu he kaituozhe* 中国林业事业的先驱和开拓者 [The pioneers of Chinese forestry] (Beijing: Zhongguo linye chubanshe, 2018), 31.

14. The collaboration between hydrologists and foresters mostly remained on the conceptual level owing to the ongoing wars. In the late 1940s, when Ling left China, the hydrologists' focus on pressing water issues outweighed the foresters' concern over soil and forests. In the socialist era, the government's plan to build dams finally drove a wedge between the hydrologists and the foresters.

15. Liang Xi 梁希, "Zheyici de chunjizaolin" 这一次的春季造林 [This time of tree planting in spring], in *Liang Xi wenji* 梁希文集 [Collected works of Liang Xi] (Beijing: Zhongguo linye chubanshe, 1983), 211.

16. Liang Xi 梁希, "Women yaoyong senlin zuowuqi he xibeidesha douzheng" 我们要用森林做武器和西北的沙斗争 [We must use forests as weapons to fight the sand in the northwest], in *Liang Xi wenji* 梁希文集 [Collected works of Liang Xi] (Beijing: Zhongguo linye chubanshe, 1983), 231–37.

17. Zhang Hanying 张含英, "Xin Huanghe de guangmingqiantu: Shuilizhuanjia Zhang Hanying zai shuili lianxihuiyishang de baogaozhaiyao" 新黄河的光明前途：水利专家张含英在水利联席会议上的报告摘要 [The bright future of the new Yellow River: Zhang Hanying's report at the joint meeting of water conservancy experts], *Renmin ribao* 人民日报, November 15, 1949.

18. Contemporary hydrologists still primarily consider silt as a significant issue

of the Yellow River. They either utilize biomineralization technologies, such as microbial-induced calcite precipitation or enzyme-induced calcite precipitation, to solidify the silt or focus on constructing hydroelectric dams. Yuke Wang et al., "Macro and Micro Experimental Study on Solidification of Yellow River Silt Based on Different Biomineralization Technologies," *Environmental Earth Sciences* 82, no. 3 (2023): 1–17; Zun Yin et al., "Irrigation, Damming, and Streamflow Fluctuations of the Yellow River," *Hydrology and Earth System Sciences* 25, no. 3 (2021): 1133–50.

19. "Shuilibu Fu Zuoyi buzhangdeng kancha huanghe xushuikuzhi wanbi" 水利部傅作义部长等勘查黄河蓄水库址完毕 [Minister Fu Zuoyi of the Ministry of Water Resources has completed a survey of the Yellow River reservoir site], *Renmin ribao* 人民日报, July 25, 1950.

20. "Ba huanghexiayou de shuihuan biancheng shuili!" 把黄河下游的水患变成水利! [Turn the flood in the lower areas of the Yellow River into water benefits!], *Renmin ribao* 人民日报, May 24, 1951.

21. Liang Xi 梁希, "Renmin zhengxie quanguoweiyuanhui disancihuiyi ershibari huishang de zhuantibaogao he fayan" 人民政协全国委员会第三次会议二十八日会上的专题报告和发言 [Special report and speech at the third session of the National Committee of the Chinese People's Political Consultative Conference on October twenty-eighth], *Renmin ribao* 人民日报, October 30, 1951.

22. Liang Xi 梁希, "Zuzhi qunzhong hulin zaolin, jianjue fandui langfei mucai" 组织群众护林造林，坚决反对浪费木材 [Organize the masses to protect forests and afforestation and resolutely oppose the waste of wood], in *Liang Xi wenji* 梁希文集 [Collected works of Liang Xi] (Beijing: Zhongguo linye chubanshe, 1983), 281–82.

23. "Sidalin pailaideren zenyangzai zhongguode heliushang gongzuozhe" 斯大林派来的人怎样在中国的河流上工作着 [How Stalin's experts work on China's rivers], *Renmin ribao* 人民日报, November 25, 1951.

24. Pietz, *Yellow River*, 153.

25. David Engerman recounts a story regarding India's technocrats that is somewhat similar to China's. He shows that during the Cold War era, some Indian experts employed the Soviet Union's expertise for assistance, while others sought out the United States to promote their own agenda. China's case is different, however, in that its foresters and hydrologists competed against each other without being fully aware that they were using knowledge drawn from two opposing camps. David Engerman, *The Price of Aid: The Economic Cold War in India* (Cambridge, MA: Harvard University Press, 2018).

26. Liang Xi 梁希, *Senlin zai guojia jingjijianshe zhong de zuoyong* 森林在国家经济建设中的作用 [The role of forests in national economic construction] (Beijing: Zhonghua quanguokexue jishu puji xiehui, 1954).

27. Stalin's plan to transform the environment was influenced and inspired by

the United States' response to the Great Depression and Dust Bowl of the 1930s. See Warde, Robin, and Sverker, *Environment*, 35.

28. Brain, *Song of the Forest*, 141.

29. For an extended discussion of the Stalin's Great Plan for the Transformation of Nature, see Brain, *Song of the Forest*, 140–67. In the early 1950s, after Stalin's successor, Nikita Khrushchev, came to power, a vast number of trees planted in Stalin's era were logged.

30. Liang, *Senlin zai guojia*, 2–3.

31. "Fourth World Forestry Congress," special issue, *Unasylva* 8, no. 3 (September 1954), https://www.fao.org/3/x5372e/x5372e00.htm.

32. Though Conrad Totman's *The Green Archipelago* mostly concerns premodern Japan, it also briefly covers modern Japanese forestry. See Conrad Totman, *The Green Archipelago: Forestry in Preindustrial Japan* (Berkeley: University of California Press, 1989), 191–214.

33. Lisa M. Brady suggests that during the 1950s, there was a conceptual shift in forestry management toward the protective functions of the forests, including combating soil erosion. Lisa M. Brady, "Sowing War, Reaping Peace: United Nations Resource Development Programs in the Republic of Korea, 1950–1953," *Journal of Asian Studies* 77, no. 2 (2018): 351–63.

34. "Woguodiyibu linye dianyingpian" 我国第一部林业电影片 [China's first forestry film], in *Zhongguo linye zhizui* 中国林业之最 [The most important issues of Chinese forestry], ed. Tang Shengxiang 汤升享 and Tang Yunli 唐云丽 (Beijing: Zhongguo linye chubanshe, 1993), 143–44.

35. The 1936 Soviet law was an expansion of a 1931 law that stressed the hydrological function of the forest. In July 1936, a new organization, the Main Administration of Forest Protection and Afforestation, was established, and a stricter law to protect the forest was implemented. Brain, *Song of the Forest*, 119.

36. Deng Zihui 邓子恢, *Guanyu genzhi huanghe shuihai he kaifahuangheshuili de zongheguihua baogao* 关于根治黄河水害和开发黄河水利的综合规划报告 [Comprehensive planning report on the radical control of the Yellow River water disaster and the development of the Yellow River water conservancy] (Beijing: Renmin chubanshe, 1955), 13.

37. "Qiannian mengxiang yaoshixian" 千年梦想要实现 [A thousand-year dream will come true], *Renmin ribao* 人民日报, July 20, 1955.

38. Zhang Hanying 张含英, *Zhiheluncong* 治河论丛 [Collections of essays on river management] (Beijing: Shangwu yinshuguan, 1936). While Zhang does very briefly mention tree planting in Yellow River management, he primarily remains focused on water management rather than on the entire natural system.

39. It is also clear that nonhuman nature and the entire planet, which Bruno Latour describes as operating as "a fully-fledged actor," contribute to the distinc-

tive modes of understanding nature. Bruno Latour, "Agency at the Time of the Anthropocene," *New Literary History* 45, no. 1 (2014): 3. For a study of nonhuman agency in modern Chinese literature, see Cheng Li and Yanjun Liu, "Reading Climate Change in the Anthropocene: Material Ecocriticism and Chinese Environmental Literature," *Green Letters* 21, no. 2 (2017): 138–51.

40. Zhonghuarenmingongheguo guowuyuan 中华人民共和国国务院, "1956 nian dao 1967 nian quanguonongye fazhan gangyao (cao'an)" 1956年到1967年全国农业发展纲要 (草案) [National agricultural development outline from 1956 to 1967 (draft)], Zhongguo zhengfuwang 中国政府网, February 6, 1956, https://www.gov.cn/gongbao/shuju/1956/gwyb195605.pdf.

41. In 1957, before the Great Leap Forward, Mao had launched the Anti-Rightist Campaign, during which the minster of the Department of Forestry Industry—an agency that was separate from the Forestry Department and was founded in 1956—was persecuted. In 1958, the two departments merged. No major changes in forestry policy occurred during or after the Anti-Rightist Campaign, showing a remarkable consistency in spite of the right-left pendulum of Maoist rhetoric.

42. "Shanxi baiwan qingshaonian kaizhan tujiyundong yigeyuenei lühua huanghefenhe" 山西百万青少年开展突击运动一月内绿化黄河汾河 [Millions of teenagers in Shanxi launch assault campaign to afforest the Fenhe River of the Yellow River within one month], *Renmin ribao* 人民日报, March 5, 1958.

43. "Quanguo zaolinyundong quanmian zhankai, Sichuan Guangxi chao'e wancheng quannian zaolinjihua" 全国造林运动全面展开四川广西超额完成全年造林计划 [Nationwide afforestation campaign fully launched, Sichuan and Guangxi exceed annual afforestation plan targets], *Renmin ribao* 人民日报, April 28, 1959. For a case study of tree planting and its local variation in Zhejiang Province, see Qiliang He, *The People's West Lake: Propaganda, Agency, and Nature in Mao's China, 1949–1976* (Honolulu: University of Hawai'i Press, 2023), 68–89.

44. *Kushan kai xianhua* 枯山开鲜花 [Barren hills blossom with fresh flowers] (Beijing: Zhonguo linye chubanshe, 1960).

45. Ma Yuping 马玉平and Wang Dewen 王得温, "Kushan kai xianhua" 枯山开鲜花 [Barren hills blossom with fresh flowers], in *Kushan kai xianhua* 枯山开鲜花 [Barren hills blossom with fresh flowers] (Beijing: Zhonguo linye chubanshe, 1960), 3.

46. Jing Long 京隆, "Wanmu 'sanbalin'" 万亩"三八林" [10,000 *mu* of "women-planted forest"], in *Kushan kai xianhua* 枯山开鲜花 [Barren hills blossom with fresh flowers] (Beijing: Zhonguo linye chubanshe, 1960), 7–15.

47. Here, gender, socialism, and labor intertwine and are reconfigured. In socialist rhetoric, women are claimed to be freed from the control of a patriarchal society. Yet planting trees partly reestablishes men's control of women because women must achieve their success under the guidance of men.

48. Again, it is the men who ultimately assess whether the planting was successful. In the evening, the women would ask men in the villages to judge their efforts. In this story, tree planting is inherently masculine, and the convergence of feminist discourse and tree-planting rhetoric makes patriarchal hierarchy resurface.

49. Zhonghuarenmingongheguo linyebu 中华人民共和国林业部, ed., *Lühua zuguo de gaochao* 绿化祖国的高潮 [The climax of greening the motherland], vol. 1 (Beijing: Beijing linye chubanshe, 1958), 22.

50. Hu Yaobang 胡耀邦, "Women you juexin lühua baotoushi" 我们有决心绿化包头市 [We have the determination to green Baotou], in *Huoyuezai lühuazuguo zhanxianshang de qingnian: Shaanxi, Gansu, Shanxi, Neimenggu, He'nan wusheng (zizhiqu) qingnian zaolindahui fayan huibian* 活跃在绿化祖国战线上的青年：陕西，甘肃，山西，内蒙古，河南五省自治区青年造林大会发言汇编 [Young people active in the greening of the motherland: compilation of speeches from the Youth Afforestation Conference in Shaanxi, Gansu, Shanxi, Inner Mongolia, and Henan Provinces and Autonomous Regions] (Beijing: Zhongguoqingnian chubanshe, 1956), 12–15.

Chapter 3

1. Hallmark examples of Mao-era ecocinema here include *Harvest* (*Fengshou* 丰收; 1953) and *Jiangnan in the North* (*Beiguo Jiangnan* 北国江南; 1963), on irrigation in northern China; *The Song of Baoshan* (*Baoshan zhige* 宝山之歌; 1958), on the exploration of iron mines during the Great Leap Forward; *Old Soldier, New Story* (*Laobing xinzhuan* 老兵新传; 1959), on wilderness transformation in Heilongjiang Province; *Daji and Her Father* (*Daji he tade fuqin* 达吉和她的父亲; 1961), on the Han Chinese construction of a dam in the ethnic Yi minority areas; *The Army's Reclamation and Battle Song* (*Junken zhange* 军垦战歌; 1965), on Xinjiang's wilderness transformation; *The Pioneers* (*Chuangye* 创业; 1974), on oilfield construction in northeastern China; and *Spring in the Desert* (*Shamo de chuntian* 沙漠的春天; 1975), on the war against the desert in Inner Mongolia.

2. The term *borderland* avoids the implied inevitability and ethnic or racial hierarchy of the concept of the *frontier*. For an examination of the concept of *borderland*, see Daniel H. Usner, "Borderlands," in *A Companion to Colonial America*, ed. Daniel Vickers (New York, NY: Blackwell, 2003), 408–24. See also Eric Taglicozzo, Helen F. Siu, and Peter C. Perdue, eds., *Asia Inside Out: Connected Places* (Cambridge, MA: Harvard University Press, 2015).

3. Alfred Crosby, *Ecological Imperialism: The Biological Expansion of Europe, 900–1900* (New York, NY: Cambridge University Press, 2004); Grove, *Green Imperialism*.

4. Grove, *Green Imperialism*, 17.

5. For excellent works on Sinophone studies and New Qing History, see Shumei Shih, "Introduction: What Is Sinophone Studies?," in *Sinophone Studies: A*

Critical Reader, ed. Shu-mei Shih, Chien-hsin Tsai, and Brian Bernards (New York, NY: Columbia University Press, 2013), 1–16; Peter C. Perdue, *China Marches West: The Qing Conquest of Central Eurasia* (Cambridge, MA: Harvard University Press, 2009).

6. Scott, *Seeing like a State*, 11–33.

7. Tsing, *Friction*, 9.

8. *Shamo de chuntian* 沙漠的春天 [Spring in the desert], directed by Zhu Wenshun 朱文顺 and Liu Zhongming 刘中明 (Changchun: Changchun Studio, 1975).

9. *Junken zhange* 军垦战歌 [The army's reclamation and battle song], directed by Yang Zhenya 杨振亚 and Deng Pu 邓普 (Beijing: August First Film Studio, 1965).

10. Zhu Kezhen 竺可桢, "Gaizao shamo shi womende lishirenwu" 改造沙漠是我们的历史任务 [Transforming the desert is our historical mission], *Renmin ribao* 人民日报, March 2, 1959.

11. "1959," Zhongguokexueyuan 中国科学院 [Chinese Academy of Sciences], accessed June 6, 2019, http://www.cas.cn/jzzky/ysss/bns/200909/t20090928_2529131.shtml.

12. Zhu Kezhen 竺可桢, "Xiangshamo jinjun" 向沙漠进军 [Marching into the desert], *Renmin ribao* 人民日报, February 9, 1961.

13. Zhu was striving not only to defeat the "enemy"—the desert—but also to harness it afterward for human use. He speculated that one day, the high winds in the desert might be used to generate electricity. And with the help of semiconductors, strong desert sunlight could also generate power, as well as manufacture salt and purify water.

14. "Lühuashamo gaizaoshamo" 绿化沙漠改造沙漠 [Greening the desert, transforming the desert], *Kexuetongbao* 科学通报 9, no. 23 (December 1958): 710–14.

15. Cheng Chongde 程崇德, "Nulishixian dadiyuanlinhuade weidalixiang" 努力实现大地园林化的伟大理想 [Striving to realize the great ideal of landscaping], *Kexuetongbao* 科学通报 10, no. 3 (March 1959): 79–81.

16. "Lühuadajun menggongshaqiu" 绿化大军猛攻沙丘 [The Green Army attacks the dunes], *Jiefangjun bao* 解放军报, November 20, 1958.

17. "Shamo lüzhou Suji tan" 沙漠绿洲苏吉滩 [Desert oasis in Suji], *Neimenggu ribao* 内蒙古日报, December 8, 1974.

18. "Wanlishamo pilüzhuang" 万里沙漠披绿装 [Covering thousands of miles of desert with green cover], *Neimenggu ribao* 内蒙古日报, October 15, 1974.

19. Neimenggu linye kexue yanjiu suo 内蒙古林业科学研究所, ed., *Neimenggu gusha zaolin shouce* 内蒙古固沙造林手册 [Inner Mongolia sand control afforestation handbook] (Beijing: Zhongguolinye chubanshe, 1960), 13–32.

20. Jing Fangrui 景芳蕊 ed., *Neimenggu zizhiqu yikezhaomeng linyezhi* 内蒙古自治区伊克昭盟林业志 [Forestry records of the Ikezhao League of the Mongolia Autonomous Region] (Hohhot: Neimenggu renmin chubanshe, 1997), 90.

21. "Shamo xinge" 沙漠新歌 [New songs from the desert], *Neimenggu ribao* 内蒙古日报, October 15, 1974.

22. Similar plotlines can be found in many films about ethnic minorities during the Mao era. For instance, in *Five Golden Flowers* (*Wu duo jin hua* 五朵金花; 1958), Bai ethnic people are represented as being "identical" with the Han in that both were oppressed by landlords and must unite to overcome their class enemy.

23. Zhu and Liu, *Shamo de chuntian.*

24. Zhu and Liu, *Shamo de chuntian.*

25. Mao Zedong 毛泽东, "Guanyu zhengque chuli renmin neibumaodun de wenti" 关于正确处理人民内部矛盾的问题 [On the correct handling of contradictions among the people], Marxists.org, February 27, 1957, https://www.marxists.org/chinese/maozedong/marxist.org-chinese-mao-19570227.htm. See also Shapiro, *Mao's War*, 1–20.

26. Zhu and Liu, *Shamo de chuntian.*

27. Zhu and Liu, *Shamo de chuntian.*

28. Anthropologist Dee Mack Williams traces the historical lineage of the Marx-Lenin-Mao line of political philosophy and its relation to ethnic environmental notions. See Dee Mack Williams, *Beyond Great Walls: Environment, Identity, and Development on the Chinese Grasslands of Inner Mongolia* (Stanford, CA: Stanford University Press, 2002).

29. Mao Tse-tung [Mao Zedong 毛泽东], "The Foolish Old Man Who Removed the Mountains," Marxists.org, June 11, 1945, revised 2004, https://www.marxists.org/reference/archive/mao/selected-works/volume-3/mswv3_26.htm.

30. Ban Wang, *The Sublime Figure of History: Aesthetics and Politics in Twentieth-Century China* (Stanford, CA: Stanford University Press, 1997).

31. Language played an important role in defining ethnicity in Mao-era China. See Thomas Mullaney, *Coming to Terms with the Nation: Ethnic Classification in Modern China* (Berkeley: University of California Press, 2011), 42–68.

32. Benedict Anderson argues that certain "imposed 'standards,'" including a unified language, are instrumental in establishing the nation-state. Thus, a film on ethnic minorities without dialect or Indigenous language contributes to the construction of a Han Chinese nation-state. Benedict Anderson, *Imagined Communities: Reflections on the Origin and Spread of Nationalism* (New York, NY: Verso Books, [1983] 2006), 81.

33. Xiaoning Lu, "The Politics of Recognition and Constructing Socialist Subjectivity: Reexamining the National Minority Film (1949–1966)," *Journal of Contemporary China* 23, no. 86 (2014): 384.

34. Zhu and Liu, *Shamo de chuntian.*

35. Bello, *Across Forest.*

36. D. Williams claims that "the pastoral Mongols have historically loved the open steppe and its spatial freedom." D. Williams, *Beyond Great Walls*, 66.

37. "Shamo lüzhou."

38. Mongolian Buddhism has been heavily influenced by Tibetan Buddhism. Vesna A. Wallace, "Buddhist Sacred Mountains, Auspicious Landscapes, and Their Agency," in *Buddhism in Mongolian History, Culture, and Society,* ed. Vesna A. Wallace (Oxford: Oxford University Press, 2015), 223.

39. D. Williams, *Beyond Great Walls,* 192.

40. D. Williams says that sand connotes freedom for Mongolians, and they respect its mobility. D. Williams, *Beyond Great Walls,* 192. During my 2015 fieldwork in Inner Mongolia, local residents claimed that there are different Mongolian vocabularies to describe sand in terms of its grain size, color, or location (such as sand found in a desert versus on a riverbank). These nuances are lost, however, as the younger generation begins to speak Mandarin.

41. Almaz Khan, "Who Are the Mongols? State, Ethnicity, and the Politics of Representation in the People's Republic of China," in *Negotiating Ethnicities in China and Taiwan,* ed. Melissa Brown (Berkeley: University of California Press, 1996), 128.

42. Zhu and Liu, *Shamo de chuntian.*

43. Mao-era China employed class dimension to exercise social control of ethnic regions. See Benno Weiner, *The Chinese Revolution on the Tibetan Frontier* (Ithaca, NY: Cornell University Press, 2020).

44. There is also a gender dimension to this clash. A Mongolian female, who embodies Communist state ideology, is at odds with a Mongolian male antagonist. During the Mao era, the replacement of Indigenous understandings of nature was portrayed as an elevation of women's social roles.

45. Tina Mai Chen, "Propagating the Propaganda Film: The Meaning of Film in CCP Writings, 1949–1965," *Modern Chinese Literature and Culture* 15, no. 2 (Fall 2003): 154–93.

46. Yomi Braester, *Painting the City Red: Chinese Cinema and the Urban Contract* (Durham, NC: Duke University Press, 2010), 6.

47. Yun Feng 云峰 and Zhuletemu 朱勒特木, "Chuntian shuyu ganyu douzheng de ren" 春天属于敢斗争的人 [Spring belongs to those who dare to fight bravely], *Renmin dianying* 人民电影 1, no. 2 (1976): 25–26. For Mao-era cinematic reception, see Jie Li, *Cinematic Guerrillas: Propaganda, Projectionists, and Audiences in Socialist China* (New York, NY: Columbia University Press, 2023).

48. Chinese Central Television (CCTV) broadcasted a six-episode version of *The Army's Reclamation and Battle Song* in 2014.

49. Joseph W. Esherick, "How the Qing Became China," in *Empire to Nation: Historical Perspectives on the Making of the Modern World,* ed. Joseph W. Esherick, Hasan Kayali, and Eric Van Young (New York, NY: Lexington Books, 2009), 248.

50. Judd Kinzley, *Natural Resources and the New Frontier: Constructing Modern China's Borderlands* (Chicago: University of Chicago Press, 2018), 99–149.

51. For Communist China's interpretation of Xinjiang's history, see the government's October 2014 white paper: State Council Information Office of the People's Republic of China, "The History and Development of the Xinjiang Production and Construction Corps," Guowuyuan xinwen bangongshi 国务院新闻办公室, October 5, 2014, http://www.scio.gov.cn/ztk/dtzt/2015/33508/33512/Document/1450269/1450269.htm.

52. Li Shenming 李慎明, "Shijiang Gebi bian Jiangnan" 誓将戈壁变江南 [Vowing to change the Gobi into Jiangnan], *Jiefangjunbao* 解放军报, October 6, 1982.

53. For excellent studies on Xinjiang's history, see David Brophy, *Uyghur Nation: Reform and Revolution on the Russia-China Frontier* (Cambridge, MA: Harvard University Press, 2016); Kinzley, *Natural Resources*; Michael Dillon, *Xinjiang: China's Muslim Far Northwest* (New York, NY: Routledge, 2003), 51–71.

54. "Haohanshahai bian lüzhou" 浩瀚沙海变绿洲 [Transforming wild desert into an oasis], *Jiefangjun bao* 解放军报, August 24, 1974.

55. Jia Yujiang 贾玉江, "Shahai tansuo" 沙海探索 [Exploring the desert], *Renmin huabao* 人民画报 25, no. 1 (1974): 35.

56. Yang and Deng, *Junken zhange*.

57. The Chinese military was mobilized to plant trees in many regions of China during the Mao era. Liang Xi celebrated the Chinese military's tree-planting effectiveness in Hubei and Nanjing. Liang Xi 梁希, "Zai 1950 nian huabei chunji zaolin zongjie huiyishangdebaogao" 在1950年华北春季造林总结会议上的报告 [Report at the 1950 spring afforestation summary meeting in northern China], in *Liang Xi wenji* 梁希文集 [Collected works of Liang Xi] (Beijing: Zhongguo linye chubanshe, 1983), 222–23.

58. "Lühua Xinjiang" 绿化新疆 [Greening Xinjiang], *Xinjiang ribao* 新疆日报, April 15, 1956.

59. "Fadong quanjiang qingshaonian kaizhan zaolinyulinhulin gongzuo" 发动全疆青少年开展造林育林护林工作 [Launch a campaign for afforestation, forest cultivation, and protection among teenagers across the entire Xinjiang region], *Xinjiang ribao* 新疆日报, March 21, 1956.

60. "Zhuajinshiji tuji qiuji lühuazaolin" 抓紧时机突击秋季绿化造林 [Seize the opportunity to carry out fall greening and afforestation], *Xinjiang ribao* 新疆日报, October 30, 1959. See also "Zizhongqu chao'e wancheng jinnian zhisha renwu" 自治区超额完成今年治沙任务 [The autonomous region overfulfilled this year's desert control tasks], *Xinjiang ribao* 新疆日报, October 30, 1959.

61. "Kaizhan jixiehua zhishuzaolin" 开展机械化植树造林 [Carry out mechanized tree planting], *Xinjiang ribao* 新疆日报, October 30, 1959.

62. Yang and Deng, *Junken zhange*.

63. Liu Mogao 刘莫皋, "*Junken zhange* paishejishi" 《军垦战歌》拍摄纪事 [The story of shooting *The army's reclamation and battle song*], *Dianyingyishu* 电影艺术 9, no. 6 (1965): 58–60.

64. Cen Shen's poem "Zoumachuan xingfeng song fengdafu chushixizheng" (走马川行奉送封大夫出师西征) depicts the harsh physical environment in the northwestern part of China in the Tang dynasty. Ts'en Shen [Cen Shen 岑参], "A Song of the Running Horse River: Presented on Saying Farewell to the Army Going on Campaign to the West," in *The Columbia Anthology of Traditional Chinese Literature*, ed. Victor Mair, trans. Daniel Bryant (New York: Columbia University Press, 2000), 101. My sincere thanks to Suiyun Pan and Kang-I Sun Chang for helping me identify this translation.

65. William Cronon, "The Trouble with Wilderness; or, Getting Back to the Wrong Nature," *Environmental History* 1, no. 1 (1996): 15.

66. Nixon, *Slow Violence*, 150.

67. Yang and Deng, *Junken zhange*.

68. Zhu Changlong 褚长龙, "Kanyingpian *Junken zhange* de ganxiang" 看影片《军垦战歌》的感想 [Reflections on watching *The army's reclamation and battle song*], *Dianyingyishu* 电影艺术 9, no. 6 (1965): 63.

69. Recent scientific research suggests that the straightforward approach of tree planting in arid regions may not be a universally suitable method for rehabilitating deteriorated dryland ecosystems. In such areas, planting trees could deplete local water supplies, enhance evapotranspiration, and potentially cause soil erosion, ultimately worsening desertification. See Qi Feng et al., "What Has Caused Desertification in China?," *Scientific Reports* 5 (2015): 15998, https://doi.org/10.1038/srep15998; Shixiong Cao, "Why Large-Scale Afforestation Efforts in China Have Failed to Solve the Desertification Problem," *Environmental Science and Technology* 42, no. 6 (March 2008): 1826–31; Changjia Li et al., "Drivers and Impacts of Changes in China's Drylands," *Nature Reviews Earth and Environment* 2, no. 12 (2021): 858–73.

70. Ramachandra Guha, *The Unquiet Woods: Ecological Change and Peasant Resistance in the Himalaya* (Berkeley: University of California Press, 1990).

71. Fedman, *Seeds of Control*, 33.

72. Karen Brown, " 'Trees, Forests and Communities': Some Historiographical Approaches to Environmental History on Africa," *Area* 35, no. 4 (2003): 343–56.

Chapter 4

1. Zhou Enlai and Jawaharlal Nehru, "Minutes of the First Meeting between Premier Zhou Enlai and Nehru," October 19, 1954, PRC FMA 204-00007-03, 14–26, Wilson Center Digital Archive, March 23, 2015, https://digitalarchive.wilsoncenter.org/document/121746.

2. Qu Bo 曲波, *Linhaixueyuan* 林海雪原 [Tracks in the snowy forest] (1957; repr., Beijing: Renmin wenxue chubanshe, 1959). For an English translation, see Chu Po [Qu Bo 曲波], *Tracks in the Snowy Forest*, trans. Sidney Shapiro (Beijing: Foreign Languages Press, 1965). Translations are mine except where otherwise noted.

3. Lin Yu 林予, "Senlin zhige" 森林之歌 [Songs of the forest], in *Senlin zhige* 森林之歌 [Songs of the forest] (Shanghai: Xinwenyi chubanshe, 1955), 93–110.

4. Within ten years after the novel was published, it was adapted as a Beijing opera (1958), a play (1958), and a film (1960). *Tracks in the Snowy Forest* was reprinted seven times from 1957 to 1958, selling approximately one million copies. Chapter 5 of the novel, which depicts the hardships of life in the forest, was revised and republished in the leading Chinese literary journal *People's Literature* (*Renmin wenxue*). See Yao Dan 姚丹, *Geming zhongguo de tongsu tezheng yu zhuti jiangou: "Linhaixueyuan" jiqi yanshengwenben kaocha* 革命中国的通俗特征与主体建构:《林海雪原》及其衍生文本考察 [Popular characteristics and main body construction of revolutionary China: a study of *Tracks in the snowy forest* and its derived texts] (Beijing: Beijing daxue chubanshe, 2011).

5. Jiang Xin, "Senlin."

6. Liang Xi 梁希, "Ru he shixian linye wunianjihua: Zai diyijie quanguo renmindaibiaodahuishangde fayan" 如何实现林业五年计划: 在第一届全国人民代表大会第二次会议上的发言 [How to realize the five-year forestry plan: a speech at the second session of the First National People's Congress], in *Liang Xi wenji* 梁希文集 [Collected works of Liang Xi] (Beijing: Zhongguo linye chubanshe, 1983), 389.

7. Sun Hui 孙辉 and Wang Jinjun 王进举, "Senlin weishi—Wujing senlinbudui de zhandoulicheng he fazhanyange" 森林卫士—武警森林部队的战斗历程和发展沿革 [Forest guardians—the combat history and development history of the People's Armed Police Forestry Force], *Junshilishi* 军事历史, no. 8 (2006): 60–64.

8. J. R. McNeill, "Woods and Warfare in World History," *Environmental History* 9, no. 3 (2004): 388–410.

9. James C. Scott, *The Art of Not Being Governed: An Anarchist History of Upland Southeast Asia* (New Haven, CT: Yale University Press, 2009).

10. Paul N. Hehn, "Serbia, Croatia and Germany 1941–1945: Civil War and Revolution in the Balkans," *Canadian Slavonic Papers* 13, no. 4 (1971): 344–73.

11. David Biggs, "Managing a Rebel Landscape: Conservation, Pioneers, and the Revolutionary Past in the U Minh Forest, Vietnam," *Environmental History* 10, no. 3 (2005): 448–76.

12. Edwin Martini, *Agent Orange: History, Science, and the Politics of Uncertainty* (Amherst: University of Massachusetts Press, 2012).

13. Qu Bo, *Linhaixueyuan*, 21.

14. William Hinton, *Fanshen: A Documentary of Revolution in a Chinese Village* (New York, NY: Monthly Review Press, 1967).

15. A novel by Zhou Libo portrays land reform as a collective effort—a political theater in which to attack and stigmatize once powerful and wealthy figures. Zhou Libo 周立波, *Baofengzhouyu* 暴风骤雨 [Hurricane] (Shijiazhuang: Huashan wenyi chubanshe, 1995).

16. Qu Bo, *Linhaixueyuan*, 32.

17. R. Harrison, *Forests*, 79.

18. Northrop Frye, *A Natural Perspective* (New York, NY: Columbia University Press, 1965), 141. See also Elizabeth Parker, *The Forest and the EcoGothic: The Deep Dark Woods in the Popular Imagination* (New York, NY: Palgrave Macmillan, 2020).

19. For a focused study of the gothic nature of the forest, see Parker, *Forest and the EcoGothic*.

20. Qu Bo, *Linhaixueyuan*, 49.

21. Qu Bo, *Linhaixueyuan*, 49.

22. Qu Bo, *Linhaixueyuan*, 82. For the translation, see Chu Po, *Tracks*, 90–91.

23. Qu Bo, *Linhaixueyuan*, 58.

24. Qu Bo, *Linhaixueyuan*, 67.

25. Qu Bo, *Linhaixueyuan*, 50–51.

26. Qu Bo, *Linhaixueyuan*, 89–90.

27. Qu Bo, *Linhaixueyuan*, 124.

28. Qu Bo, *Linhaixueyuan*, 464. I mostly use S. Shapiro's translation here. See Chu Po, *Tracks*, 459.

29. Qu Bo, *Linhaixueyuan*, 465.

30. Qu Bo, *Linhaixueyuan*, 370.

31. Qu Bo, *Linhaixueyuan*, 493.

32. Qu Bo, *Linhaixueyuan*, 493–94.

33. Qu Bo, *Linhaixueyuan*, 188.

34. Lin, "Senlin zhige," 99.

35. For the bildungsroman in the Mao era, see Mingwei Song, *Young China: National Rejuvenation and the Bildungsroman, 1900–1959* (Cambridge, MA: Harvard University Asia Center, 2015), 286–333.

36. Xiaobing Tang, *Visual Culture in Contemporary China* (Cambridge: Cambridge University Press, 2015), 18–60; Zhu Yu 朱羽, *Shehuizhuyi yu "ziran": 1950–1960 niandai Zhongguo meixue lunzheng yu wenyi shijian yanjiu* 社会主义与"自然" : 1950–1960年代中国美学论争与文艺实践研究 [Socialism and "nature": study of aesthetic debates and artistic practices in China during the 1950s–1960s] (Beijing: Beijing daxue chubanshe, 2020), 27–59.

37. Lin, "Senlin zhige," 96.

38. Lin, "Senlin zhige," 96.

39. The discourse of love plays a pivotal role in the nationalist-oriented political and social imagination of twentieth-century China. Haiyan Lee argues that a revolutionary feeling has contributed to identity formation for the Chinese people as a connected community with shared goals. Lee, *Revolution of the Heart*, 222. Similarly, Jianmei Liu suggests that "revolution plus love" negotiates the intensifying

conflicts between the individual and the collective and the private and the public. Jianmei Liu, *Revolution Plus Love: Literary History, Women's Bodies, and the Thematic Repetition in Twentieth-Century Chinese Fiction* (Honolulu: University of Hawai'i Press, 2003), 19. See also ibid., 1–32. While Lee's and Liu's discussions of revolution plus love concern relations among human beings, this thought is readily applicable to "Senlin zhige." In the Mao era, the nation-state intentionally cultivated affection for the natural world.

40. Lin, "Senlin zhige," 96–97.

41. Lin, "Senlin zhige," 97–98.

42. Lin, "Senlin zhige," 97–98.

43. Lin, "Senlin zhige," 99.

44. Lin, "Senlin zhige," 101.

45. Lin, "Senlin zhige," 101.

46. Lin, "Senlin zhige," 107–8.

47. Elizabeth Perry, "Moving the Masses: Emotion Work in the Chinese Revolution," *Mobilization: An International Quarterly* 7, no. 2 (2002): 111–28.

48. Lin, "Senlin zhige," 110.

49. The country of Malaya consisted of the states on the Malay Peninsula. In 1963, Malaya joined with Singapore, Sarawak, and North Borneo to form the Federation of Malaysia.

Sinophone studies have explored the role of languages and the natural world in understanding local community and indigeneity outside of mainland China. Carlos Rojas, "Language, Ethnicity, and the Politics of Literary Taxonomy: Ng Kim Chew and Mahua Literature," *PMLA* 131, no. 5 (2016): 1316–27; Brian Bernards, *Writing the South Seas: Imagining the Nanyang in Chinese and Southeast Asian Postcolonial Literature* (Seattle: University of Washington Press, 2015), 109–35. For the globality of Mahua literature, see Cheow Thia Chan, *Malaysian Crossings: Place and Language in the Worlding of Modern Chinese Literature* (New York, NY: Columbia University Press, 2022).

50. Ng Kim Chew 黄锦树, *Nanyang renmingongheguo beiwanglu* 南洋人民共和国备忘录 [Memorandums of the South Seas people's republic] (Taipei: Lianjing chubanshe, 2013); Jin Zhimang 金枝芒, *Ji'e* 饥饿 [Starvation] (Kuala Lumpur: Ershiyi shiji chubanshe, 2008). I thank David Der-wei Wang for recommending this latter novel to me.

51. It is important to note that support for the MCP was not unanimous among the Chinese in Malaya. The British colonial strategy of creating new villages to isolate rural populations from the guerrillas, combined with improved economic conditions and political reforms, gradually eroded support for the MCP. Additionally, the MCP's rigid adherence to Communist ideology and its association with Communist China also alienated some segments of the population, particularly as

the Cold War intensified. By the late 1950s and 1960s, the MCP's influence had significantly waned, and it never gained substantial support from the Malay majority or the Indian minority in Malaya. The Malayan Emergency officially ended in 1960, although the MCP continued a lower-level insurgency into the 1980s. For a history of decolonization in Malaysia and its cultural politics, see Alison M. Groppe, *Sinophone Malaysian Literature: Not Made in China* (Amherst, MA: Cambria Press, 2013), 25–102.

52. Ng, *Nanyang*, 112–13.

53. Jin, *Ji'e*, 5.

54. Jin, *Ji'e*, 65.

55. Jin, *Ji'e*, 123.

56. Jin, *Ji'e*, 28.

57. Jin, *Ji'e*, 58.

58. Jin, *Ji'e*, 214.

59. Edward O. Wilson, "Biophilia and the Conservation Rhetoric," in *The Biophilia Hypothesis*, ed. Stephen R. Kellert and Edward O. Wilson (Washington, DC: Island Press, 1995), 32.

60. Simon C. Estok, *The Ecophobia Hypothesis* (New York, NY: Routledge, 2018).

61. Jin, *Ji'e*, 149.

62. Jin, *Ji'e*, 219.

63. Studying the relationship between environmental values and geographical locations, Yi-fu Tuan argues that "attachment to place can emerge paradoxically from the experience of nature's intransigence." For example, marginal farmers in the Great Plains of the United States have intimate ties to the land and even develop pride in their ability to endure. Yi-Fu Tuan, *Topophilia: A Study of Environmental Perceptions, Attitudes, and Values* (New York, NY: Columbia University Press, 1974), 97.

64. Jin, *Ji'e*, 366.

65. Nicolai Volland, *Socialist Cosmopolitanism: The Chinese Literary Universe, 1945–1965* (New York, NY: Columbia University Press, 2017), 113. For a summary of sci-fi literary history in modern Chinese literature, see Mingwei Song, *Fear of Seeing: A Poetics of Chinese Science Fiction* (New York, NY: Columbia University Press, 2023), 105–15.

66. Tao Bennai 陶本艿, "Qihougongsi de gushi" 气候公司的故事 [The story of the Climate Company], in *Zhongguo kehuanxiaoshuo daquan* 中国科幻小说大全 [The complete collection of Chinese science fiction novels], ed. Rao Zhonghua 饶忠华 (Beijing: Haiyang chubanshe, 1982), 99–100.

67. Liu Xingshi 刘兴诗, "Beifang de yun" 北方的云 [The cloud of the north], in *Zhongguo kehuanxiaoshuo daquan* 中国科幻小说大全 [The complete collection of

Chinese science fiction novels), ed. Rao Zhonghua 饶忠华 (Beijing: Haiyang chubanshe, 1982), 133–34.

68. In the United States, sci-fi was tinged with environmental concerns in the 1970s. Published in 1972, Ursula K. Le Guin's *The Word for World Is Forest* depicts the revolt of the Indigenous population of the fictional planet Athshe after it is colonized by human beings from Earth. The colonizers, having depleted Earth's natural resources, deforest Athshe to create farmland. In the Athshean language, the word for *forest* is the same as that for *world,* showing the dependence of the Athshean culture on the forest. The novel shows that the viciousness of the humans is linked to their disconnection from the forest. Ursula K. Le Guin, *The Word for World Is Forest* (New York, NY: Tor, 1972). See also Chris Pak, *Terraforming: Ecopolitical Transformations and Environmentalism in Science Fiction* (Liverpool: Liverpool University Press, 2016), 132–34; Elizabeth Cummins, *Understanding Ursula K. Le Guin* (Columbia: University of South Carolina Press, 1990), 98–103.

69. The comic book adaptation of the story by Mao Liangying and Zhu Guangrong was published in 1980. Yang Xunan 杨叙男 and Tao Jinsheng 陶锦生, "Milin shaobing" 密林哨兵 [Scout in the jungle], *Shaonian kexue* 少年科学 4, no. 4 (1979): 12–22. For the picture-book (*lianhuanhua*) adaptation, see Mao Liangying 毛亮英, *Milin shaobing* 密林哨兵 [Scout in the jungle], illus. Zhu Guangrong 朱光荣 (Lanzhou: Gansu renmin chubanshe, 1980).

70. *Fighting in the Jungle* is perhaps derived from a novella of the same name, published in 1959. The novella describes the Communists' fight against the Chinese collaborator with the Japanese in the forests of northeastern China during the Second World War. Zhang Ge 张革, *Milin li de zhandou* 密林里的战斗 [Fighting in the jungle] (Shanghai: Shaonianertong chubanshe, 1959).

71. Xiao Jianheng 肖建亨, "Milin hu zong" 密林虎踪 [Tiger traces in the forest], *Shaonian kexue* 少年科学 2, no. 4 (1977): 13–24.

72. Xiao Jianheng, "Milin hu zong," 53.

73. The depiction of the animal refuge in "Tiger Traces in the Forest" reflects the protective dimension of the Communist Chinese government in the Mao era. In 1956, Dinghushan National Nature Reserve in Guangdong Province became the first nature reserve in China; as of 2023, there were 2,729 such reserves, covering nearly 15 percent of the country's land area. The creation of reserves embodies the state's assertion of authority over and ownership of Chinese wildlands and animal species. Panda protection was highly successful even during the Cultural Revolution. Zhijun Ma et al., "Changes in Area and Number of Nature Reserves in China," *Conservation Biology* 33, no. 5 (2019): 1066–75; Songster, *Panda Nation*, 52–83.

Chapter 5

1. For an excellent study of the social and political history of 1980s China, see Dingxin Zhao, *The Power of Tiananmen: State-Society Relations and the 1989 Beijing Student Movement* (Chicago: University of Chicago Press, 2004), 39–78. For cultural dynamics in the 1980s, see Jing Wang, *High Culture Fever: Politics, Aesthetics, and Ideology in Deng's China* (Berkeley: University of California Press, 1996); Jason McGrath, *Postsocialist Modernity: Chinese Cinema, Literature, and Criticism in the Market Age* (Stanford, CA: Stanford University Press, 2008); Cheng Guangwei 程光炜, *Chongfan bashi niandai* 重返八十年代 [Revisiting the 1980s] (Beijing: Beijingdaxue chubanshe, 2009).

2. For an environmental history of China in the 1980s, see Economy, *River Runs Black*, 59–90; Robert Marks, *China: Its Environment and History* (New York, NY: Lexington Books, 2011), 265–330.

3. Ouyang, Li, and Ouyang, *Heisedang'an*, 213.

4. Xu Gang 徐刚, *Famuzhe, xinglai!* 伐木者,醒来! [Woodcutter, wake up!] (Changchun: Jilin renmin chubanshe, 1997); Ah Cheng [阿城], *The King of Trees*, in *The King of Trees*, trans. Bonnie S. McDougall (New York, NY: New Directions, 2010), 1–56.

5. Xu Gang 徐刚, *Guonan* 国难 [National disaster] (Beijing: Zuojia chubahsen, 2005); Xu Gang 徐刚, *Diqiu zhuan* 地球传 [Biography of the earth] (Beijing: Zuojia chubanshe), 2009; Xu Gang 徐刚, *Da senlin* 大森林 [Big forest] (Beijing: Beijing shiyue wenyi chubanshe, 2017); Xu Gang 徐刚, "Ziran biji" 自然笔记 [Notebook of nature], *Renmin wenxue* 人民文学 74, no. 1 (2022): 4–21. Some details were confirmed with the author in 2019 and 2022.

6. Li Guo sheds light on *Woodcutter, Wake Up!* by means of a binary opposition between biocentric and anthropocentric environmental ideas. Li Guo, "From the Margins to the Hearth," *Modern Chinese Literature and Culture* 31, no. 2 (2019): 162–206.

7. Xu Gang, *Famuzhe*, 1. *Woodcutter, Wake Up!* provoked discussion of Chinese forestry and the natural environment in the late 1980s.

8. Yue Meng examines how the notion of *ecology* is associated with social Darwinism and monocrop farming. Here, I focus on the scientific trajectory of ecology. See Yue Meng, "Unspeakable Ecology: Eco-science and Environmental Awareness through Thick Inquiries, 1910s–1980s," *Twentieth-Century China* 47, no. 2 (2022): 91–111. See also Stevan Harrell, *An Ecological History of Modern China* (Seattle: University of Washington Press, 2023).

9. Donald Worster, *Nature's Economy: A History of Ecological Ideas* (Cambridge: Cambridge University Press, 1994), 192. In English, the word *economy*, which originally referred to the behavior of households, shares a similar etymology with *ecology*, which refers to "the household behavior" of living organisms. This seems an

apt metaphor, and earlier scholars employed "the economy of nature" to describe the interactions of living organisms. It was not until the end of nineteenth century that the term *ecology* gained traction.

10. Sharon E. Kingsland, *The Evolution of American Ecology, 1890–2000* (Baltimore: Johns Hopkins University Press, 2005), 183–87.

11. Worster, *Nature's Economy*, 203.

12. Tansley pushes back against the idea that a natural system is made up of random associations that change according to the laws of probability. In addition, he suggests that humans are an essential part of the system. Arthur G. Tansley, "The Use and Abuse of Vegetational Concepts and Terms," *Ecology* 16, no. 3 (1935): 284–307.

13. Kingsland, *Evolution of American Ecology*, 185. See also Frank B. Golley, *A History of the Ecosystem Concept in Ecology: More than the Sum of the Parts* (New Haven, CT: Yale University Press, 1993).

14. G. Evelyn Hutchinson, "Circular Causal Systems in Ecology," *Annals of the New York Academy of Sciences* 50 (1948): 221–46.

15. Eugene P. Odum, *Fundamentals of Ecology* (Philadelphia, PA: Saunders, 1953); Howard T. Odum, *Environment, Power, and Society* (New York, NY: Wiley, 1971); Eugene P. Odum, "The Ecosystem Approach in the Teaching of Ecology Illustrated with Sample Class Data," *Ecology* 38 (1957): 531–35.

16. Zhang Ting 张珽 and Dong Shuangqiu 董爽秋, *Zhiwu shengtaixue* 植物生态学 [Plant ecology] (Guangzhou: Guangzhou weixing yinshuachang, 1930), 7.

17. Fei Hongnian 费鸿年, *Dongwu shengtaixue* 动物生态学 [Animal ecology] (Shanghai: Shangwu yinshuguan, 1933).

18. Fei, *Dongwu shengtaixue*, 35–36.

19. Yano Munemoto 矢野宗幹, *Kunchong shengtaixue* 昆虫生态学 [Insect ecology], trans. Xue Deyu 薛德焴 (Beijing: Shangwu yinshuguan, 1935).

20. Nakano Harufusa 中野治房, *Zhiwu qunluoxue xiaoyin* 植物群落学小引 [A short introduction to the plant community], trans. Yu Jingliang 于景良 (Beijing: Shangwu yinshuguan, 1936).

21. Fei Hongnian 费鸿年, *Dongwu shengtaixue gangyao* 动物生态学纲要 [Animal ecology outline] (Shanghai: Zhonghuashuju, 1937).

22. Huang Xiuming 黄修明, *Kunchong shengtaixue gailun* 昆虫生态学概论 [Introduction to insect ecology] (Shanghai: Zhonghuashuju, 1941).

23. Qu Zhongxiang 曲仲湘, Wen Zhenwang 文振旺, and Zhu Kegui 朱克贵, "Nanjing linggusi senlin xianzhuangde fenxi" 南京灵谷寺森林现况的分析 [Analysis of the current state of the Linggu Temple Forest in Nanjing], *Zhiwu xuebao* 植物学报 1 (1952): 18–49.

24. Le Tianyu 乐天宇, "Zhiwu shengtai xianghu zuoyong guanxi de jiben guilü" 植物生态相互作用关系的基本规律 [Basic principles of plant ecological interactions], *Linye kexue* 林业科学 3 (1956): 185–213.

25. The Chinese translation of Shennikov's *Plant Ecology* was published in 1953. A. P. Shennikov, *Zhiwu shengtaixue* 植物生态学 [Plant ecology], trans. Wang Wen (Shanghai: Xinnongchubanshe, 1953).

26. Le Tianyu, "Zhiwu shengtai." Le's role in China's environmental science was tainted by his faith in Lysenkoism, which rejects natural selection and genetics in favor of the inheritance of acquired characteristics. Nevertheless, his updated system of ecology brought the idea of understanding nature as an interactive system to Chinese environmental science. For Lysenkoism's impacts in China, see Laurence Schneider, "Lysenkoism in China: Proceedings of the 1956 Qingdao Genetics Symposium," *Chinese Law and Government* 19, no. 2 (Summer 1986): iii–xxii.

27. In 1957, Qian Xuesen published the seminal article "Engineering Cybernetics" ("Gongcheng kongzhi lun"), which was based on his 1954 English-language monograph *Engineering Cybernetics*, in the Chinese magazine *Popular Science* (*Kexue dazhong*). Educated at MIT and The California Institute of Technology in the 1930s, Qian worked for the Manhattan Project in the 1940s and returned to China in 1955. He played a pivotal role in various Chinese technological breakthroughs: in addition to his efforts to advance Chinese atomic and hydrogen bomb and missile programs, he popularized cybernetics. In the article, he explains that a crucial factor of cybernetics is creating feedback. Just as humans must observe their surroundings to adjust their actions, machines need such mechanisms to adapt to changing environments. According to Qian, a system that can adjust and control itself automatically is the most desirable; one of the primary goals in the field of cybernetics is to transform unstable systems into more-stable ones. He later drafted articles in various fields, including military studies, agriculture, information systems, and social science. Qian Xuesen 钱学森, "Gongcheng kongzhi lun" 工程控制论 [Engineering cybernetics], in *Qian Xuesen wenji* 钱学森文集 [Collections of Qian Xuesen], vol. 1, ed. Gu Jihuan 顾吉环, Li Ming 李明, and Tu Yuanji 涂元季 (Beijing: Guofang gongye chubanshe, 2012), 229. Qian's articles on cybernetics began to appear in the early 1950s in the newspaper *Guangming Daily* (*Guangming ribao*) and elsewhere. See also Qian Xuesen 钱学森, *Lun xitonggongcheng* 论系统工程 [On cybernetics] (Changsha: Hunan kexuejishu chubanshe, 1982).

28. Hu Shizhi 胡式之, "Zhiwu shengtaixue dizhiwuxue diyici xueshuhuiyi" 植物生态学地植物学第一次学术会议 [First academic conference on plant ecology and geobotany], *Kexue tongbao* 科学通报8, no. 1 (1963): 63.

29. Qu Geping 曲格平, *Huanjing juexing: Renlei huanjinghuiyi he Zhongguo diyici huanjingbaohu huiyi* 环境觉醒：人类环境会议和中国第一次环境保护会议 [Environmental awakening: the Conference on the Human Environment and China's first environmental protection conference] (Beijing: Zhongguo huanjing kexue chubanshe, 2010), 5.

30. Qu Geping, *Huanjing juexing*, 60.

31. Xu Gang, interview by author, July 20, 2019.

32. Trees and forests received increasing attention in 1980s literary works. First published in Heilongjiang Province in the early 1980s, the magazine *Forestry Literature* (*Linye wenxue*) documents the protective function of trees in the environment, which is reminiscent of the writings of foresters examined in chapters 1 and 2. An article entitled "Lishi yu dierdai senlin" (History and second-generation forests) quotes a speech by Liang Xi suggesting that the knowledge of forestry had begun to take hold among grassroots groups. The article argues that "green mountains exist forever, and then economic management sustains forever" (qingshan chang zai, yong xu jingying). According to this vision, trees are planted for human use. The article's articulation of sustainable development laid the groundwork for later texts like *Woodcutter, Wake Up!* The government's dramatic shift in the 1980s from depicting Ma Yongshun as a model logger to showing him as a tree-planting model worker reflects the rise of conservationist rhetoric in the post-Mao era. Working on a forest farm in northeastern China in 1959, Ma received praise from Premier Zhou Enlai for his efforts at tree harvesting and its application to industrial construction. In the 1980s, as the forestry resources dwindled, he became increasingly aware of the need to plant trees. He then transformed himself from a tree-cutting model worker to an icon of tree planting. See Wang Xiaolin 王晓琳, "Lishi yu dierdai senlin" 林业与第二代森林 [History and second-generation forests], *Linye wenxue* 林业文学 3, no. 3 (June 1989): 11–17; Wu Baosan 吴宝三 and Cao Feng 曹锋, *Ma Yongshun zhuan* 马永顺传 [Biography of Ma Yongshun] (Harbin: Heilongjiang renmin chubanshe, 1999).

33. Founded in Rome, Italy, in 1968, the Club of Rome published the influential *The Limits to Growth* in 1972 to foster global environmental activism. See John Robert McNeill and Peter Engelke, *The Great Acceleration: An Environmental History of the Anthropocene since 1945* (Cambridge, MA: Harvard University Press, 2016), 188.

34. Xu, *Famuzhe, xinglai!*, 2.

35. Xu, *Famuzhe, xinglai!*, 80.

36. Xu, *Famuzhe, xinglai!*, 3.

37. Xu, *Famuzhe, xinglai!*, 75–76.

38. The story of John Muir and Gifford Pinchot is more complicated than Xu makes it out to be. Muir advocated for preservation, while Pinchot promoted conservation. For Muir, the wilderness was a place to be revered without the intrusion of humankind. He saw foresters and other conservationists as meddling intruders into nature's world. Pinchot, though agreeing on the need to protect the natural environment, also believed that public wilderness areas could be used as a source of income for the country if the resources were handled wisely. Whether intentional or not, the simplification of US environmental history in Xu's work reflects a desire to use the West as a prime example for China's environmental movement.

See Donald Worster, *A Passion for Nature: The Life of John Muir* (New York, NY: Oxford University Press, 2008); John Meyer, "Gifford Pinchot, John Muir, and the Boundaries of Politics in American Thought," *Polity* 30, no. 2 (1997): 267–84. For a nuanced study of Pinchot, especially his transformation during the final decade of his life, see C. Miller, *Gifford Pinchot.*

39. Nash, *Wilderness*, 108–21.

40. Xu Gang, interview.

41. Katherine Reischl, *Photographic Literacy: Cameras in the Hands of Russian Authors* (Ithaca, NY: Cornell University Press, 2018), 101–46.

42. Ursula Heise, *Sense of Place and Sense of Planet: The Environmental Imagination of the Global* (New York, NY: Oxford University Press, 2008), 13.

43. Wu, *Empires of Coal*, 129–59. See also Kenneth Pomeranz, *The Great Divergence: China, Europe, and the Making of the Modern World Economy* (Princeton, NJ: Princeton University Press, 2000).

44. Wang Jinxi, "the Iron Man," first put forward these words, which were set to music in songs that circulated in China to encourage sacrifice, adherence to Communist ideology, and self-discipline. See Hou Li, *Building for Oil: Daqing and the Formation of the Chinese Socialist State* (Cambridge, MA: Harvard University Asia Center, 2018), 148.

45. Xu Gang's other works also situate Chinese environmental problems within a global context. His *Green Proclamation* (*Lüse xuanyan*), for instance, begins with the fate of planet Earth, admonishes the Chinese to treat the natural world with a caring and humble attitude, and suggests that a campaign against desertification in China is a crusade for all human beings. Xu's 2017 Lu Xun Literary Prize–winning masterpiece, *Big Forest* (*Da senlin*), is a panoramic depiction of the Chinese woodlands. Going back to the advent of *Homo sapiens*, *Big Forest* elaborates on the ecological function of the forest from the perspective of the *longue durée*. Forests give birth to a lively world when organisms begin to emerge and eventually lead to the appearance of human beings. Xu Gang 徐刚, *Lüse xuanyan* 绿色宣言 [Green proclamation] (Changchun: Shidai wenyi chubanshe, 1997); Xu Gang, *Da senlin*.

46. *Woodcutter, Wake Up!* embodies an intellectual trend in the 1980s of exposing Chinese environmental degradation and expanding it to a broader scale. For instance, Sha Qing's 1986 work of reportage "Beijing shiqupingheng" documents water shortages in the capital resulting from industrial development and misuse of natural resources. It situates China's environmental problems alongside such incidents as the Great Smog of London, the Love Canal in the United States, deforestation in the Soviet Union, and the Minamata incident in Japan and urges readers to consider China from a global ecological perspective. Sha Qing 沙青, "Beijing shiqupingheng" 北京失去平衡 [Beijing out of balance], *Xinhuawenzhai* 新华文摘 12, no. 6 (1986): 123–36. For a concise examination of post-1980 representation of trees

in modern Chinese literature, see Weijie Song, "Environmentality, Sustainability, and Chinese Storytelling," *Cultura* 20, no. 1 (2023): 62–63.

47. Xu, *Famuzhe, xinglai!*, 38.

48. Almost at the same time in the United States, the dumping of PCB-contaminated soil in a predominately Black community in North Carolina triggered massive protests, leading to a humanistic turn with respect to environmental issues—what scholars now call *environmental justice*.

49. Xu's critical insight represents what Charles Laughlin calls "the aesthetics of historical experience," which is embodied by the transformation of reportage from a literary genre into a communication medium during the 1980s. Charles Laughlin, *Chinese Reportage: The Aesthetics of Historical Experience* (Durham, NC: Duke University Press, 2002), 265.

50. Xiaomei Chen, *Occidentalism: A Theory of Counter-Discourse in Post-Mao China* (Oxford: Oxford University Press, 1995), 27–48.

51. The *New Observer* (*Xin guancha*) and *World Economic Herald* (*Shijie jingji daobao*) were two leading liberal, or pro-West, media outlets in China during the 1980s. The shutting down of the latter in Shanghai by then mayor Jiang Zemin contributed to his rise to general secretary of the Communist Party in 1989.

52. "*Famuzhe, xinglai!* zuotanjiyao" 《伐木者，醒来!》座谈纪要 [*Woodcutter, wake up!* minutes of the symposium], *Xin guancha* 新观察 20, no. 6 (1988): 2–6.

53. Zhongguo huanjingnianjian bianji weiyuanhui 中国环境年鉴编辑委员会, ed., *Zhongguo huanjingnianjian 1992* 中国环境年鉴1992 [China environment yearbook 1992] (Beijing: Zhongguo huanjingkexue chubanshe, 1992), 217. "Huanjing wenxue yanjiuhui zaijingchengli" 环境文学研究会在京成立 [Environmental Literature Research Association established in Beijing], *Renmin ribao* 人民日报, March 25, 1991.

54. Wang Meng 王蒙, "Zanmei lüye" 赞美绿叶 [Praising green leaves], *Lüye* 绿叶 1, no. 1 (1992): 7.

55. To reduce urban unemployment, promote Maoist ideals among the educated youth, and bridge the urban-rural divide, Mao launched the Down to the Countryside movement in the 1950s, "sending down" urban youths to rural areas. It significantly impacted the "sent-down youth," reshaping China's social fabric. Approximately seventeen million youths went to rural areas during the campaign. Ah Cheng, whose real name is Zhong Acheng, attended a highly regarded secondary school in Beijing prior to being dispatched to the countryside during the Cultural Revolution. In Xishuangbanna, the educated youths felled more than six hundred thousand acres, even changing the region's climate. Thornber, *Ecoambiguity*, 146–47. Emily Honing and Xiaojian Zhao detail the extensive and complex negotiations between rural and urban officials during this movement. Emily Honig and Xiaojian Zhao, *Across the Great Divide: The Sent-Down Youth Movement in Mao's China, 1968–1980* (New York, NY: Cambridge University Press, 2019).

56. For an insightful analysis of roots-seeking literature and its cultural back-

ground of sent-down youths, see He Guimei 贺桂梅, *"Xinqimeng" zhishidangan: Bashiniandai Zhongguo wenhua yanjiu* "新启蒙"知识档案：八十年代中国文化研究 ["New Enlightenment" knowledge archives: Chinese cultural studies in the 1980s] (Beijing: Beijing daxue chubanshe, 2010), 164–219.

57. Han Shaogong 韩少功, "Wenxue de gen" 文学的根 [The roots of literature], in *Zai houtai de houtai* 在后台的后台 [In the background of the background] (Beijing: Renmin wenxue chubanshe, 2008), 273. The article was originally published in *Writer* (*Zuojia*) magazine in 1985. The term *searching for roots* was first used by cultural critic Li Tuo in an essay that appeared in early 1984, a year before Han's proclamation. Scholars also trace the seeds of root-seeking literature by Wang Zengqi, Wang Meng, and Jia Pingwa in the early 1980s. See Kuang Xinnian 旷新年, *Xiezai dangdai wenxue bianshang* 写在当代文学边上 [Written on the edge of contemporary literature] (Shanghai: Shanghai jiaoyu chubanshe, 2005), 59–76.

58. Han elaborates, "The Chu culture clearly differs from that of 'descendants of the dragon' along the banks of the Yellow River. People of Chu are 'descendants of the bird,' who worship, extol, and imitate birds." Han Shaogong, "Wenxue de gen," 273.

59. The scope of roots-seeking literature is broader than its beginnings as a simple revival of Chu culture would indicate. Writers from different locations emphasize their own local cultures in revitalizing Chinese literature as well as engaging with the West. Mo Yan's story about Shandong villages, Zhang Chengzhi's depiction of Inner Mongolia grasslands, and Wang Zengqi's portrayal of lyrical rural life in Beijing represent disparate but similar attempts to negotiate between probing Chinese local tradition and interacting with the Western world. For a review of roots-seeking literature and its underlying dynamics, see Hong Zicheng 洪子诚, *Zhongguo wenxue 1949–1989* 中国文学1949–1989 [Chinese literature 1949–1989] (Beijing: Beijing chubanshe, 2020), 217–34. For an excellent English-language summary of roots-seeking literature, see Mark Leenhouts, "Culture against Politics: Roots-Seeking Literature," in *The Columbia Companion to Modern Chinese Literature*, ed. Kirk Denton (New York, NY: Columbia University Press, 2016), 299–306. See also Cheng Guangwei, *Chongfan*.

60. Representatives of pseudomodernists include Xu Xing and Liu Suola. Huang Ziping 黄子平, "Guanyu 'weixiandaipai' jiqi piping" 关于"伪现代派"及其批评 [On "pseudomodernism" and its criticism], *Beijing wenxue* 北京文学 2 (1988): 4; Leenhouts, "Culture against Politics," 299.

61. Ah Cheng [阿城], *The King of Chess*, in *The King of Trees*, trans. Bonnie S. McDougall (New York, NY: New Directions, 2010), 57–120. See also Ah Cheng [阿城], *The King of Children*, in *The King of Trees*, trans. Bonnie S. McDougall (New York, NY: New Directions, 2010), 123–78.

62. Chen Xiaoming 陈晓明, *Biaoyi de jiaolü* 表意的焦虑 [The anxiety of expression] (Beijing: Zhongyang bianyi chubanshe, 2004), 81.

63. Meng Li and King-fai Tam, "Literature of Trauma and Reflection," in *Routledge Handbook of Modern Chinese Literature*, ed. Ming Dong Gu (New York, NY: Routledge, 2019), 450–61.

64. Ah, *King of Trees*, 32.

65. Ah, *King of Trees*, 22–23.

66. Ah, *King of Trees*, 32.

67. A muntjac is a small deer, also known as the barking deer or rib-faced deer, which is native to South and Southeast Asia. Note mine.

68. Ah, *King of Trees*, 9–11.

69. For discussions of the allusion to the late fourth-century BCE Chinese philosopher Zhuangzi, see Visser, *Questioning Borders*, 41; Thornber, *Ecoambiguity*, 511–12.

70. Ah, *King of Trees*, 13.

71. Ah, *King of Trees*, 43.

72. Gang Yue, "The Strange Landscape of the Ancient: Environmental Consciousness in '*The King of Trees*,'" *American Journal of Chinese Studies* 5, no. 1 (1998): 68–88; Thornber, *Ecoambiguity*, 147–55.

73. Sigmund Freud distinguishes between *mourning* as a finished act of grieving and *melancholia* as an uncompleted mourning that displaces itself onto other items. Sigmund Freud, "Mourning and Melancholia," *Journal of Nervous and Mental Disease* 56, no. 5 (1922): 543–45.

74. These emotions, according to Heise, blend melancholia, desire, ignorance, and political impulse for the irretrievable past. Ursula Heise, *Imagining Extinction: The Cultural Meanings of Endangered Species* (Chicago: University of Chicago Press, 2016), 34–35. Judith Butler identifies the power of mourning in nation building. After the terrorist attacks on September 11, 2001, in the United States, public mourning became a process of recovering from excruciating trauma. Judith Butler, *Precarious Life: The Powers of Mourning and Violence* (New York, NY: Verso Books, 2006). According to sociologist Catriona Mortimer-Sandilands, "Melancholia is not only a denial of the loss of a beloved object, but also a potentially politicized way of preserving that object in the midst of a culture that fails to recognize its significance." Catriona Mortimer-Sandilands, "Melancholy Natures, Queer Ecologies," in *Queer Ecologies: Sex, Nature, Politics, Desire*, ed. Catriona Mortimer-Sandilands and Bruce Erikson (Bloomington: Indiana University Press, 2010), 331–58.

75. Ah, *King of Trees*, 49.

76. Ah, *King of Trees*, 55.

77. Ah, *King of Trees*, 55–56.

78. Ah, *King of Trees*, 56.

79. While *The King of Trees* shows that a person can become integrated with a

tree, a lesser-known Ah Cheng short story, "Shuzhuang," shows a tree seemingly becoming animate and resembling a person. At the beginning of the story, a stump stands as the remains of an old tree in a small town, where men and women sing songs and lead peaceful lives. But the Cultural Revolution interferes, and they are forced to stop singing. After the Cultural Revolution is over, the residents realize that the tree stump is actually a famous singer, who, unlike the younger generation, knows how to sing old songs. Tree Stump exchanges songs with an old woman, singing that he is an old tree. The transformation of Tree Stump from the remains of the tree to a person is reminiscent of *The King of Trees*. "Shuzhuang" also reveals that the folk culture of singing songs can animate a tree. Ah Cheng 阿城, "Shuzhuang" 树桩 [Tree stump], *Renmin wenxue* 人民文学 23, no. 10 (1984): 67–69.

Coda

1. Wang Junling 王俊岭, "5.5yishuang jiao 'zhong'chule zhepian 'mayisenlin' " 5.5亿双脚"种"出了这片"蚂蚁森林" [550 million pairs of feet "planted" this "ant forest"], *Renminwang* 人民网, June 29, 2020, http://paper.people.com.cn/rmrbhwb/html/2020-06/29/content_1994419.htm.

2. "Mayin jinrongfuwujituan zhengshi chengli, tuozhan jinrong xinshengtai" 蚂蚁金融服务集团正式成立，拓展金融新生态 [Ant Financial Services Group was formally established to expand the new financial ecology], Alibaba Group, October 16, 2014, https://www.alibabagroup.com/cn/news/article?news=p141016. Accessed December 10, 2020, the link is no longer working as of April 23, 2024.

3. In her study of computer games, Alenda Chang argues that instead of alienating users from nature, games can build a potent connection with the natural world while establishing an equally significant relationship to technology. See Alenda Chang, *Playing Nature: Ecology in Video Games* (Minneapolis: University of Minnesota Press, 2019).

4. Max Horkheimer and Theodor W. Adorno, *Dialect of Enlightenment: Philosophical Fragments*, trans. Edmund Jephcott (Stanford, CA: Stanford University Press, 2002), 115–16.

5. For games and the emotional mechanisms underlying them, see Patrick Jagoda, *Experimental Games: Critique, Play, and Design in the Age of Gamification* (Chicago: University of Chicago Press, 2020), 153–90.

6. "Nidui zhifubao *Mayisenlin* youshenme kanfa?" 你对支付宝《蚂蚁森林》有什么看法? [What do you think of Alipay *Ant Forest*?], Zhihu 知乎, August 30, 2016, https://www.zhihu.com/question/50162658; "Ruhe pingjia zhifubao *Mayisenlin* 3nian zhong zhenshu 1yi ke, weishenme neng xiyin zhemeduoren yiqi zhongshu?" 如何评价支付宝《蚂蚁森林》3年种真树1亿棵，为什么能吸引这么多人一起种树? [How do you evaluate that Alipay *Ant Forest* has planted 100 million real trees in three

years, why can it attract so many people to plant trees together?], Zhihu 知乎, April 22, 2019, https://www.zhihu.com/question/321188877.

7. Zizi Papacharissi, *Affective Publics: Sentiment, Technology, and Politics* (New York, NY: Oxford University Press, 2015), 12.

8. "Ruhe pingjia."

9. *Ant Forest* resembles its competitor Tencent's earlier product, *QQ Farm*, a social-network game that encourages its users to grow their own crops and steal those of other players. The similarity of the two products reflects Alibaba's desire to lure users from its competitor. For a study of Chinese social-media dynamics and competition, see Jing Wang, *The Other Digital China: Nonconfrontational Activism on the Social Web* (Cambridge, MA: Harvard University Press, 2019), 96–135.

10. There is, of course, a confrontational dimension to Chinese environmental activism. Studying the opposition to building dams on the Nu River, Guobin Yang and Craig Calhoun contend that there is a rising green public sphere of critical environmental discourse in China in mass media and on the internet. As we can see from the example of *Ant Forest*, nonconfrontational environmental activism is gaining traction in China as it begins to promote the interests of both the masses and the ruling party. Guobin Yang and Craig Calhoun, "Media, Civil Society, and the Rise of a Green Public Sphere in China," *China Information* 21, no. 2 (July 2007): 211–36. See also Jing Wang, *Other Digital China*, 132.

11. Nixon, *Slow Violence*, 2.

12. "Nidui zhifubao."

13. For a case study of Starbucks's greenwashing, see Tiffany Derville Gallicano, "A Critical Analysis of Greenwashing Claims," *Public Relations Journal* 5, no. 3 (2011): 1–21.

14. See Davis Harper, "Starbucks Falls Short on Environmental Commitments: Reports Suggest That Caffeine Lovers' Iconic Mermaid May Not Be So Green After All," *Sierra*, November 21, 2017, https://www.sierraclub.org/sierra/starbucks-falls-short-environmental-commitments. For an examination of greenwashing, see Daniel C. Esty and Andrew Winston, *Green to Gold: How Smart Companies Use Environmental Strategy to Innovate, Create Value, and Build Competitive Advantage* (New Haven, CT: Yale University Press, 2006).

15. For various definitions of consumer environmentalism, see Monika Stoma and Agnieszka Dudziak, "Analysis of Market Behaviour on the Organic Food Market in Terms of Environmental Protection and Consumer Environmentalism," in *International Symposium on Farm Machinery and Processes Management in Sustainable Agriculture*, ed. Simone Pascuzzi and Francesco Santoro (Cham: Springer, 2022), 348.

16. Peter Dauvergne, *Environmentalism of the Rich* (Cambridge, MA: MIT Press, 2016), 14.

17. Dauvergne, *Environmentalism of the Rich*, 14.

18. Lisa Rafel, *Desiring China: Experiments in Neoliberalism, Sexuality, and Public Culture* (Durham, NC: Duke University Press, 2007), 6.

19. B. Wang, *Sublime Figure of History*, 15.

20. For a historical study of the dialectical relationship between the individual and the crowd in the Chinese cultural imagination, see Tie Xiao, *Revolutionary Waves: The Crowd in Modern China* (Cambridge, MA: Harvard University Press, 2017).

21. "Zhifubaode *Mayisenlin* shi zhende dui shengtai chansheng le bangzhume, haishi zhishi biaomian gongfu?" 支付宝的《蚂蚁森林》是真的对生态产生了帮助么，还是只是做表面功夫? [Does Alipay's *Ant Forest* really help the ecology, or is it just a superficial effort?], Zhihu 知乎, March 6, 2018, https://www.zhihu.com/question/268277647. As of December 27, 2020, the top response has received 52,700 likes—ten times higher than the criticism of *Ant Forest*.

22. "Ruhe pingjia."

23. "The Paris Agreement," United Nations, accessed March 22, 2023, https://unfccc.int/process-and-meetings/the-paris-agreement/the-paris-agreement.

24. "Zhongguode mayijinfu zhanshi le ruhe tongguo shuzi yingxiangli yingdui qihoubianhua" 中国的蚂蚁金服展示了如何通过数字影响力应对气候变化 [China's Ant Financial demonstrates how to tackle climate change through digital influence], Lianheguohuanjingguihuashu 联合国环境规划署, November 3, 2017, https://www.unenvironment.org/zh-hans/xinwenyuziyuan/gushi/zhongguodemayijinfuzhanshileruhetongguoshuziyingxiangliyingduiqihoubianhua.

25. When smog covered almost half of China's territory in 2017, discussions regarding the validity of the Green Wall emerged. Arguments circulated that trees prevent the wind from blowing the smog away. In fact, trees do not stop the 1,500-meter-high wind from blowing away the particles, and the leaves of trees can absorb these particles as well. While trees have little relation to the formation of smog, this incident indirectly shows how the Three-North Shelter Forest program is far from transparent to the general public. See You Xueqing 游雪晴, "Sanbei fanghulin dangzhu fengchuiwumai? Zhuanjia: Shu bubeizhe 'guo'" 三北防护林挡住风吹雾霾? 专家: 树不背这"锅" [Does the Three-North Shelter belt block the wind and smog? Expert: trees should be the "scapegoat"], Huanqiuwang 环球网, January 18, 2017, https://tech.huanqiu.com/article/9CaKrnJZQjl. See also Zhai et al., "Assessing the Effects," 1–12.

26. Timothy Mitchell, *Carbon Democracy: Political Power in the Age of Oil* (New York, NY: Verso Books, 2011), 1–42.

27. Rachel Carson Papers, box 83, Beinecke Rare Book and Manuscript Library, Yale University.

28. Chris Buckley, "Chinese Doctor, Silenced after Warning of Outbreak, Dies from Coronavirus," *New York Times*, February 6, 2020.

29. Emily T. Yeh, "The Making of Natural Infrastructure in China's Era of Ecological Civilization," *China Quarterly* 255 (September 2023): 611–27.

30. Other authoritarian regimes have also used conservationist initiatives for political purposes. In the Soviet Union, Joseph Stalin launched his Great Plan for the Transformation of Nature in 1948. It aimed to show the ideological superiority of Marxism over capitalism through environmental conservation. In Zimbabwe, Robert Mugabe leveraged the social importance of environmental issues to win support. Acknowledging their efforts provincializes the West and its environmentalism in a liberal democratic background. Stephen Brain, "Stalin's Environmentalism," *Russian Review* 69, no. 1 (2010): 93–118; Muchaparara Musemwa, "Zimbabwe's Drift toward Authoritarianism and Its Environmental Consequences, 2000–2017," in *Environmentalism under Authoritarian Regimes: Myth, Propaganda, Reality*, ed. Stephen Brain and Viktor Pal (New York, NY: Routledge, 2019), 191–213.

31. For China's environmental impacts in Africa in Chinese cultural representations, see Cajetan Iheka, *African Ecomedia: Network Forms, Planetary Politics* (Durham, NC: Duke University Press, 2021), 205–12.

32. In 2022, e.g., China continued to build coal-fueled power plants in Turkey and central Asia. As Victor Seow observes, there is a "contradiction between concurrent domestic decarbonization and carbon exportation." Victor Seow, *Carbon Technocracy: Energy Regimes in Modern East Asia* (Chicago: University of Chicago Press, 2022), 314. See also Dan Murtaugh, "China Has Built 14 Overseas Coal Plants since Vowing No New Ones," *Bloomberg*, September 22, 2022, https://www.bloomberg.com/news/articles/2022-09-22/china-has-built-14-overseas-coal-plants-since-vowing-no-new-ones.

BIBLIOGRAPHY

Primary Sources

Ah Cheng [阿城]. *The King of Chess*. In *The King of Trees*, translated by Bonnie S. McDougall, 57–120. New York, NY: New Directions, 2010.

———. *The King of Children*. In *The King of Trees*, translated by Bonnie S. McDougall, 123–78. New York, NY: New Directions, 2010.

———. *The King of Trees*. In *The King of Trees*, translated by Bonnie S. McDougall, 1–56. New York, NY: New Directions, 2010.

Ah Cheng 阿城. *Shuwang* 树王 [The king of trees]. In *Qiwang, Shuwang, Haiziwang* 棋王，树王，孩子王, 71–130. Taipei: Dadi Chubanshe, 2007.

———. "Shuzhuang" 树桩 [Tree stump]. *Renmin wenxue* 人民文学 23, no. 10 (1984): 67–69.

Alibaba. *Ant Forest*. V. 10.2.10.7020. Alibaba, 2020. iOS 14 or later.

Alumni Records Office. Beinecke Rare Book and Manuscript Library, Yale University.

An Shinong 安事农. *Zhongguo senlinfa* 中国森林法 [Chinese forestry law]. Shanghai: Huatong shuju, 1933.

"Ba huanghexiayou de shuihuan biancheng shuili!" 把黄河下游的水患变成水利! [Turn the flood in the lower areas of the Yellow River into water benefits!]. *Renmin ribao* 人民日报, May 24, 1951.

Bailie, Joseph. "Famine and Afforestation in China." *International Review of Mission* 6, no. 3 (1917): 450–56.

———. Papers. Yale Divinity School Archive.

———. "Magna Charta of China's Forestry Work." *American Forestry* 22 (May 1916): 268–72.

Bao Shen 葆琛. "Zhishujie de laili" 植树节的来历 [The origin of Arbor Day]. *Nongmin* 农民2, no. 4 (1926): 4–6.

Buckley, Chris. "Chinese Doctor, Silenced after Warning of Outbreak, Dies from Coronavirus." *New York Times*, February 6, 2020.

Carson, Rachel. Papers. Beinecke Rare Book and Manuscript Library, Yale University.

Chen Bochui 陈伯吹, Lü Jinlu 吕金录, and Zhuang Yu 庄俞, eds. *Fuxing guoyu keben* 复兴国语课本 [Reviving the Chinese-language textbook]. Vol. 5. Shanghai: Shangwu yinshuguan, 1935.

Chen Xiaoyuan 陈晓原. "Xinzhongguo de senlin shiye" 新中国的森林事业 [The forestry industry in new China]. *Renmin ribao* 人民日报, September 15, 1950.

Cheng Chongde 程崇德. "Nulishixian dadiyuanlinhuade weidalixiang" 努力实现大地园林化的伟大理想 [Striving to realize the great ideal of landscaping]. *Kexuetongbao* 科学通报 10, no. 3 (March 1959): 79–81.

"Cheng dazongtongwen: Niding qingming wei zhishujie qingyi shenling xuanshi" 呈大总统文: 拟定清明为植树节请以申令宣示 [To the president: proposing Qingming as Arbor Day, please declare by presidential order]. *Nongshang gongbao* 农商公报 2, no. 1 (1915): 7.

Chi Xinhong 池昕鸿, ed. *Chen Guofu quanzhuan* 陈果夫全传 [Complete works of Chen Guofu]. Yanji: Yanbian renmin chubanshe, 2003.

Chu Po [Qu Bo 曲波]. *Tracks in the Snowy Forest*. Translated by Sidney Shapiro. Beijing: Foreign Languages Press, 1965.

Dai Qing 戴晴, ed. *Changjiang, Changjiang: Sanxia gongcheng lunzheng* 长江,长江: 三峡工程论争 [Yangtze, Yangtze: the Three Gorges project debate]. Guiyang: Guizhou renmin chubanshe, 1989.

"Dazongtong junjian guowuyuan neiwu jiaoyunongshang gezongzhang junjian bennian zhishujie yejing dianling daoxian gexuexiao yili juxing zhishu jinian" 大总统钧鉴国务院内务教育农商各总长钧鉴: 本年植树节业经电令道县各学校一例举行植树纪念 [To the president's august attention, to the state council's ministers of internal affairs, education, agriculture, and commerce: this year's Arbor Day activities have been ordered by telegram to be uniformly conducted as a commemorative event in schools at all levels across the prefectures and counties]. *Zhengfu gongbao* 政府公报 450 (1917): 21.

Deng Zihui 邓子恢. *Guanyu genzhi huanghe shuihai he kaifahuangheshuili de zongheguihua baogao* 关于根治黄河水害和开发黄河水利的综合规划报告 [Comprehensive planning report on the radical control of the Yellow River water disaster and the development of the Yellow River water conservancy]. Beijing: Renmin chubanshe, 1955.

Department of Forestry Catalogue. Manuscripts and Archives, Yale University Library.

"Fadong quanjiang qingshaonian kaizhan zaolinyulinhulin gongzuo" 发动全疆青少年开展造林育林护林工作 [Launch a campaign for afforestation, forest cultivation, and protection among teenagers across the entire Xinjiang region]. *Xinjiang ribao* 新疆日报, March 21, 1956.

"*Famuzhe, xinglai!* zuotanjiyao" 《伐木者，醒来!》座谈纪要 [*Woodcutter, wake up!* minutes of the symposium]. *Xin guancha* 新观察 20, no. 6 (1988): 2–6.

Fei Hongnian 费鸿年. *Dongwu shengtaixue* 动物生态学 [Animal ecology]. Shanghai: Shangwu yinshuguan, 1933.

———. *Dongwu shengtaixue gangyao* 动物生态学纲要 [Animal ecology outline]. Shanghai: Zhonghuashuju, 1937.

"Fourth World Forestry Congress." Special issue, *Unasylva* 8, no. 3 (September 1954). https://www.fao.org/3/x5372e/x5372e00.htm.

"Gaiding chunfen wei zhishujie" 改定春分为植树节 [Change Arbor Day to the spring equinox]. *Guoli disanzhongshandaxue jiaoyu zhoukan* 国立第三中山大学教育周刊 21 (1928): 5.

Gao Chunxian 高春先. "Gucheng zhongxin xiaoxue 3.12 zhishujie huodong baodao" 古城中心小学3.12植树节活动报道 [Report on the Arbor Day activity at Gucheng Central Primary School on March 12]. *Nanjingshi Gucheng xiaoxue* 南京市古城小学, March 13, 2013. http://www.njsxd.org/news_detail.asp?id=308965.

"Gebu jinian sunzongli, tongrij uxing zhishujie" 各埠纪念孙总理，同日举行植树节 [Various towns commemorate Premier Sun, hold Arbor Day on the same day]. *Shishixinbao (Shanghai)* 时事新报(上海), March 13, 1929.

Guan Bingzong 关秉宗. "Zhishu huodong fansi: Que yongxu linye kanshu kandao feizhouqu" 植树活动省思：缺永续林业砍树砍到非洲去 [Reflections on tree-planting activities: lacking sustainable forestry, cutting down trees extends to Africa]. *Lianhebao* 联合报, March 14, 2016.

Guan Zhong 管仲. *Guanzi* 管子 [Writings of Master Guan]. Edited by Li Yuanyan 李元燕 and Li Wenjuan 李文娟. Guangzhou: Guangzhou chubanshe, 2001.

Guoyu keben 国语课本 [Mandarin textbook]. Vol. 3. N.p.: Huabeixinhuashudian, 1948.

Han An 韩安. "Shijie geguo guoyou senlin dashi" 世界各国国有森林大势 [The state of state-owned forests in various countries worldwide]. *Nonglingongbao* 农林公报 2, no. 3 (1913): 135–43.

———. "Shijie geguo guoyou senlin dashi" 世界各国国有森林大势 [The state of state-owned forests in various countries worldwide]. *Nonglingongbao* 农林公报 2, no. 4 (1913): 93–98.

———. "Shijie geguo guoyou senlin dashi" 世界各国国有森林大势 [The state of state-owned forests in various countries worldwide]. *Nonglingongbao* 农林公报 2, no. 5 (1913): 131–38.

"Han juxing zhishu dianli" 汉举行植树典礼 [Hankou holds tree-planting ceremony]. *Shenbao* 申报, March 13, 1943.

Han Shaogong 韩少功. "Wenxue de gen" 文学的根 [The roots of literature]. In *Zai houtai de houtai* 在后台的后台 [In the background of the background], 273–79. Beijing: Renmin wenxue chubanshe, 2008.

Hao Jingsheng 郝景盛. *Senlin wannenglun* 森林万能论 [The omnipotence of forests]. Shanghai: Zhengzhong shuju, 1947.

———. "Zhishujie shuguan" 植树节书感 [Reflections on Arbor Day]. *Dagongbao* 大公报, March 12, 1948.

"Haohanshahai bian lüzhou" 浩瀚沙海变绿洲 [Transforming wild desert into an oasis]. *Jiefangjun bao* 解放军报, August 24, 1974.

Hongse famuzhe 红色伐木者 [Red loggers]. Beijing: Zhonguo linye chubanshe, 1960.

Hu Jie'an 胡洁庵. "Zhishu ji" 植树记 [Documenting tree planting]. *Anhui jiaoyu* 安徽教育 2, no. 11 (1931): 64–65.

Hu Shizhi 胡式之. "Zhiwu shengtaixue dizhiwuxue diyici xueshuhuiyi" 植物生态学地植物学第一次学术会议 [First academic conference on plant ecology and geobotany]. *Kexue tongbao* 科学通报8, no. 1 (1963): 63.

Hu Yaobang 胡耀邦. "Women you juexin lühua baotoushi" 我们有决心绿化包头市 [We have the determination to green Baotou]. In *Huoyuezai lühuazuguo zhanxianshang de qingnian: Shaanxi, Gansu, Shanxi, Neimenggu, He'nan wusheng (zizhiqu) qingnian zaolindahui fayan huibian* 活跃在绿化祖国战线上的青年：陕西，甘肃，山西，内蒙古，河南五省自治区青年造林大会发言汇编 [Young people active in the greening of the motherland: compilation of speeches from the Youth Afforestation Conference in Shaanxi, Gansu, Shanxi, Inner Mongolia, and Henan Provinces and Autonomous Regions], 3–22. Beijing: Zhongguoqingnian chubanshe, 1956.

Huang Lide 黄立德 and Zeng Aidi 曾艾狄. "Zhishu de fangfa" 植树的方法 [Methods of planting trees]. *Jiefang ribao* 解放日报, April 5, 1943.

Huang Taiming 黄泰明. "Lun Qingming zhishu shifou xiangyi" 论清明植树是否相宜 [On whether it is appropriate to plant trees during the Qingming festival]. *Xuesheng wenyi congkan* 学生文艺丛刊 3, no. 6 (1926): 49–50.

Huang Xiuming 黄修明. *Kunchong shengtaixue gailun* 昆虫生态学概论 [Introduction to insect ecology]. Shanghai: Zhonghuashuju, 1941.

Huang Yingdai 黄影呆. "Zhishujie ganyan" 植树节感言 [Arbor Day speech]. *Xinwen bao* 新闻报, March 14, 1933.

"Huanjing wenxue yanjiuhui zaijingchengli" 环境文学研究会在京成立 [Environmental Literature Research Association established in Beijing]. *Renmin ribao* 人民日报, March 25, 1991.

"Huichang fandongfenzi fanghuoshaoshan" 会昌反动分子放火烧山 [Huichang reactionaries set fire to the mountain]. *Hongse Zhonghua* 红色中华, no. 177 (1934): 2.

Jia Yujiang 贾玉江. "Shahai tansuo" 沙海探索 [Exploring the desert]. *Renmin huabao* 人民画报 25, no. 1 (1974): 35.

Jiang Jieshi [Chiang Kai-shek] 蒋介石. "Xin shenghuo yundong zhi yaoyi" 新生活运动之要义 [Key ideas of the New Life movement]. In *Xin shenghuo yundong* 新生活运动 [The New Life movement], edited by Ye Chucang 叶楚伧, 8–23. Nanjing: Zhongzheng shuju, 1934.

Jiang Jieshi zhishu 蒋介石植树 [Chiang Kai-shek plants trees]. March 12, 1929. Photograph. Minguotupian ziyuanku 民国图片资源库, Zhenjiang, Jiangsu Province.

Jiang Xin 江心. "Senlin de jianjie zuoyong" 森林的间接作用 [Indirect benefits of forests]. *Jiefang ribao* 解放日报, September 26, 1944.

———. "Senlin yousha yongchu" 森林有啥用处 [What the use of forests is]. *Jiefang ribao* 解放日报, September 24, 1944.

———. "Yan'anshi zhishu diaocha" 延安市植树调查 [Tree-planting survey of Yan'an]. *Jiefang ribao* 解放日报, March 2, 1945.

———. "Zhishu he hulin" 植树和护林 [Tree planting and forest protection]. *Jiefang ribao* 解放日报, March 25, 1942.

Jin Zhimang 金枝芒. *Ji'e* 饥饿 [Starvation]. Kuala Lumpur: Ershiyi shiji chubanshe, 2008.

Jing Fangrui 景芳蕊, ed. *Neimenggu zizhiqu yikezhaomeng linyezhi* 内蒙古自治区伊克昭盟林业志 [Forestry records of the Ikezhao League of Mongolia Autonomous Region]. Hohhot: Neimenggu renmin chubanshe, 1997.

Jing Long 京隆. "Wanmu 'sanbalin'" 万亩"三八林" [10,000 *mu* of "women-planted forest"]. In *Kushan kai xianhua* 枯山开鲜花 [Barren hills blossom with fresh flowers], 7–16. Beijing: Zhonguo linye chubanshe, 1960.

"Jingxuesheng canbai guofu tongxiang" 京学生参拜国父铜像 [Students in Nanjing pay homage to the bronze statue of the founding father]. *Shenbao* 申报, March 13, 1943.

"Jinri qingming bianqugedi qunzhong zhunbei daliangzhishu" 今日清明，边区各地群众准备大量植树 [Today is the Qingming festival, and people from all over the border area are preparing to plant a large number of trees]. *Jiefang ribao* 解放日报, April 5, 1946.

"Jushokusai ni atsumatta utsukushii hassen no hitobito, ame no fuzei mo hitoiori fusawashiku, kyou Hibiya no seikan" 樹植祭に集まった美しい８０００の人々：雨の風情も一入ふさわしくけふ日比谷の盛観 [The beautiful 8,000 people gathered at Arbor Day: the rain added a fitting atmosphere to today's grand spectacle at Hibiya]. *Asahi shinbun* 朝日新聞, April 4, 1928.

"Kaizhan jixiehua zhishuzaolin" 开展机械化植树造林 [Carry out mechanized tree planting]. *Xinjiang ribao* 新疆日报, October 30, 1959.

Kushan kai xianhua 枯山开鲜花 [Barren hills blossoms fresh flowers]. Beijing: Zhonguo linye chubanshe, 1960.

Le Guin, Ursula K. *The Word for World Is Forest*. New York, NY: Tor, 1972.
Le Tianyu 乐天宇. "Zhiwu shengtai xianghu zuoyong guanxi de jiben guilü" 植物生态相互作用关系的基本规律 [Basic principles of plant ecological interactions]. *Linye kexue* 林业科学3 (1956): 185–213.
Li Shenming 李慎明. "Shijiang Gebi bian Jiangnan" 誓将戈壁变江南 [Vowing to change the Gobi into Jiangnan]. *Jiefangjunbao* 解放军报, October 6, 1982.
Liang Xi 梁希. "Minshengwenti yu senlin" 民生问题与森林 [The problems of livelihood and the forest]. In *Liang Xi wenji* 梁希文集 [Collected works of Liang Xi], 1–7. Beijing: Zhongguo linye chubanshe, 1983.
———. "Renmin zhengxie quanguoweiyuanhui disancihuiyi ershibari huishang de zhuantibaogao he fayan" 人民政协全国委员会第三次会议二十八日会上的专题报告和发言 [Special report and speech at the third session of the National Committee of the Chinese People's Political Consultative Conference on October twenty-eighth]. *Renmin ribao* 人民日报, October 30, 1951.
———. "Ru he shixian linye wunianjihua: Zai diyijie quanguo renmindaibiaodahuishangde fayan" 如何实现林业五年计划: 在第一届全国人民代表大会第二次会议上的发言 [How to realize the five-year forestry plan: a speech at the second session of the First National People's Congress]. In *Liang Xi wenji* 梁希文集 [Collected works of Liang Xi], 388–94. Beijing: Zhongguo linye chubanshe, 1983.
———. *Senlin zai guojia jingjijianshe zhong de zuoyong* 森林在国家经济建设中的作用 [The role of forests in national economic construction]. Beijing: Zhonghua quanguokexue jishu puji xiehui, 1954.
———. "Women yaoyong senlin zuowuqi he xibeidesha douzheng" 我们要用森林做武器和西北的沙斗争 [We must use forests as weapons to fight the sand in the northwest]. In *Liang Xi wenji* 梁希文集 [Collected works of Liang Xi], 231–37. Beijing: Zhongguo linye chubanshe, 1983.
———. "Zai 1950nian huabei chunji zaolin zongjie huiyishangdebaogao" 在1950年华北春季造林总结会议上的报告 [Report at the 1950 spring afforestation summary meeting in northern China]. In *Liang Xi wenji* 梁希文集 [Collected works of Liang Xi], 217–30. Beijing: Zhongguo linye chubanshe, 1983.
———. "Zheyici de chunjizaolin" 这一次的春季造林 [This time of tree planting in spring]. In *Liang Xi wenji* 梁希文集 [Collected works of Liang Xi], 210–14. Beijing: Zhongguo linye chubanshe, 1983.
———. "*Zhongguo linye* fakanci" 《中国林业》发刊词 [Foreword to *Chinese forestry*]. In *Liang Xi wenji* 梁希文集 [Collected works of Liang Xi], 207–9. Beijing: Zhongguo linye chubanshe, 1983.
———. "Zuzhi qunzhong hulin zaolin, jianjue fandui langfei mucai" 组织群众护林造林，坚决反对浪费木材 [Organize the masses to protect forests and afforestation and resolutely oppose the waste of wood]. In *Liang Xi wenji* 梁希文集 [Collected works of Liang Xi], 279–83. Beijing: Zhongguo linye chubanshe, 1983.

Lin, Dau-yang [Ling Daoyang 凌道扬]. *Chapters on China and Forestry*. Shanghai: Commercial Press, 1916.

Lin Yu 林予. "Senlin zhige" 森林之歌 [Songs of the forest]. In *Senlin zhige* 森林之歌 [Songs of the forest], 93–110. Shanghai: Xinwenyi chubanshe, 1955.

Ling Daoyang 凌道扬. *Ling Daoyang quanji* 凌道扬全集 [Complete works of Ling Daoyang]. Edited by Liu Zhongguo 刘中国. Hong Kong: Gongyuan chuban-youxiangongsi, 2009.

———. *Senlin yaolan* 森林要览 [Forest outline]. Shanghai: Shangwu yinshuguan, 1918.

———. *Senlinxue dayi* 森林学大意 [Main ideas of forestry]. Shanghai: Shangwu yinshuguan, 1916.

"Lingmu qian zhishu dianli" 陵墓前植树典礼 [Tree-planting ceremony in front of the mausoleum]. *Minguo ribao* 民国日报, March 13, 1929.

"Linshi zhongyangzhengfu wengao: Renmin weiyuanhui duiyu zhishuyundongde jueyian" 临时中央政府文告: 人民委员会对于植树运动的决议案 [Provisional central government announcement: resolution of the People's Committee on the Tree-Planting Movement]. *Hongse Zhonghua* 红色中华, no. 15 (1932): 6.

Liu Mogao 刘莫皋. "*Junken zhange* paishejishi" 《军垦战歌》拍摄纪事 [The story of shooting *The army's reclamation and battle song*]. *Dianyingyishu* 电影艺术 9, no. 6 (1965): 58–60.

Liu Wenhui 刘文辉. "Dayuejin zhongde linye jianshe linyebu buzhang liuwenhui de fayan" 大跃进中的林业建设林业部部长刘文辉的发言 [Forestry construction during the Great Leap Forward: speeches by forestry minister Liu Wenhui]. *Renmin ribao* 人民日报, April 6, 1960.

Liu Xingshi 刘兴诗. "Beifang de yun" 北方的云 [The cloud of the north]. In *Zhongguo kehuanxiaoshuo daquan* 中国科幻小说大全 [The complete collection of Chinese science fiction novels], edited by Rao Zhonghua 饶忠华, 133–34. Beijing: Haiyang chubanshe, 1982.

Liu Zongyuan 柳宗元. "Zhongshu Guo Tuotuo zhuan" 种树郭橐驼传 [The legend of Guo Tuotuo planting trees]. In *Zhongguo gudai linye wenxian xuandu* 中国古代林业文献选读 [Selected readings from ancient Chinese forestry literature], edited by Zhang Lianwei 张连伟, Li Fei 李飞, and Zhou Jingyong 周景勇, 196–99. Beijing: Beijing yanshan chubanshe, 2015.

"Lühua Xinjiang" 绿化新疆 [Greening Xinjiang]. *Xinjiang ribao* 新疆日报, April 15, 1956.

"Lühuadajun menggongshaqiu" 绿化大军猛攻沙丘 [The Green Army attacks the dunes]. *Jiefangjun bao* 解放军报, November 20, 1958.

"Lühuashamo gaizaoshamo" 绿化沙漠改造沙漠 [Greening the desert, transforming the desert]. *Kexuetongbao* 科学通报 9, no. 23 (December 1958): 710–14.

Ma Ganlin 马甘霖. "Zhishujie" 植树节 [Arbor Day]. *Ertong shijie (Shanghai 1922)* 儿童世界 (上海1922) 34, no. 5 (1935): 68–71.

Ma Yuping 马玉平and Wang Dewen 王得温. "Kushan kai xianhua" 枯山开鲜花 [Barren hills blossom with fresh flowers]. In *Kushan kai xianhua* 枯山开鲜花 [Barren hills blossom with fresh flowers], 1–6. Beijing: Zhonguo linye chubanshe, 1960.

Mao Liangying 毛亮英. *Milin shaobing* 密林哨兵 [Scout in the jungle]. Illustrated by Zhu Guangrong 朱光荣. Lanzhou: Gansu renmin chubanshe, 1980.

Mao Tse-tung [Mao Zedong 毛泽东]. "The Foolish Old Man Who Removed the Mountains." Marxists.org, June 11, 1945. Revised 2004. https://www.marxists.org/reference/archive/mao/selected-works/volume-3/mswv3_26.htm.

Mao Zedong 毛泽东. "Guanyu zhengque chuli renmin neibu maodun de wenti" 关于正确处理人民内部矛盾的问题 [On the correct handling of contradictions among the people]. Marxists.org, February 27, 1957. https://www.marxists.org/chinese/maozedong/marxist.org-chinese-mao-19570227.htm.

"Mayin jinrongfuwu jituan zhengshi chengli, tuozhan jinrong xinshengtai" 蚂蚁金融服务集团正式成立，拓展金融新生态 [Ant Financial Services Group was formally established to expand the new financial ecology]. Alibaba Group, October 16, 2014. https://www.alibabagroup.com/cn/news/article?news=p141016.

Meng Fan 梦梵, ed. *Shijing* 诗经 [Book of songs]. Beijing: Tuanjie chubanshe, 1995.

"Menglie kaizhan guangda zhishu yundong: Wei tigao nongyeshengchan er douzheng, zenyang qu jinxingzhishu yundong" 猛烈开展广大植树运动：为提高农业生产而斗争，怎样去进行植树运动 [Vigorously launch a wide-scale tree-planting campaign: strive to improve agricultural production, how to conduct a tree-planting movement]. *Hongse Zhonghua* 红色中华, no. 145 (1934): 2.

"Minguo sanshiqinian zhishujie teji: Nonglinbu zuobuzhangshunsheng zhishujie zhuci" 民国三十七年植树节特辑：农林部左部长舜生植树节祝词 [The Republic of China's thirty-seventh year Arbor Day special: Arbor Day message by Minister Zuo Shunsheng of the Ministry of Agriculture and Forestry]. *Linchantongxun* 林产通讯 2, no. 6 (1948): 2.

Muller, Charles. "Mencius (Selections)." Resources for East Asian Language and Thought. Accessed March 12, 2021. http://www.acmuller.net/con-daoenciuss.html.

Murtaugh, Dan. "China Has Built 14 Overseas Coal Plants since Vowing No New Ones." *Bloomberg*, September 22, 2022. https://www.bloomberg.com/news/articles/2022-09-22/china-has-built-14-overseas-coal-plants-since-vowing-no-new-ones.

Nakano Harufusa 中野治房. *Zhiwu qunluoxue xiaoyin* 植物群落学小引 [A short introduction to the plant community]. Translated by Yu Jingliang 于景良. Beijing: Shangwu yinshuguan, 1936.

Neimenggu linye kexue yanjiu suo 内蒙古林业科学研究所, ed. *Neimenggu gusha zaolin shouce* 内蒙古固沙造林手册 [Inner Mongolia sand control afforestation handbook]. Beijing: Zhongguolinye chubanshe, 1960.

Ng Kim Chew 黄锦树. *Nanyang renmingongheguo beiwanglu* 南洋人民共和国备忘录 [Memorandums of the South Seas people's republic]. Taipei: Lianjing chubanshe, 2013.

"Nidui zhifubao *Mayisenlin* youshen mekanfa?" 你对支付宝《蚂蚁森林》有什么看法? [What do you think of Alipay *Ant Forest*?]. Zhihu 知乎, August 30, 2016. https://www.zhihu.com/question/50162658.

Qian Xuesen 钱学森. "Gongcheng kongzhi lun" 工程控制论 [Engineering cybernetics]. In *Qian Xuesen wenji* 钱学森文集 [Collections of Qian Xuesen], vol. 1, edited by Gu Jihuan 顾吉环, Li Ming 李明, and Tu Yuanji 涂元季, 227–30. Beijing: Guofang gongye chubanshe, 2012.

———. *Lun xitonggongcheng* 论系统工程 [On cybernetics]. Changsha: Hunan kexuejishu chubanshe, 1982.

"Qiannian mengxiang yaoshixian" 千年梦想要实现 [A thousand-year dream will come true]. *Renmin ribao* 人民日报, July 20, 1955.

Qu Bo 曲波. *Linhaixueyuan* 林海雪原 [Tracks in the snowy forest]. 1957. Reprint, Beijing: Renmin wenxue chubanshe, 1959.

Qu Geping 曲格平. *Huanjing juexing: Renlei huanjinghuiyi he Zhongguo diyici huanjingbaohu huiyi* 环境觉醒: 人类环境会议和中国第一次环境保护会议 [Environmental awakening: the Conference on the Human Environment and China's first environmental protection conference]. Beijing: Zhongguo huanjing kexue chubanshe, 2010.

Qu Zhongxiang 曲仲湘, Wen Zhenwang 文振旺, and Zhu Kegui 朱克贵. "Nanjing linggusi senlin xianzhuang fenxi" 南京灵谷寺森林现况的分析 [Analysis of the current state of the Linggu Temple forest in Nanjing]. *Zhiwu xuebao* 植物学报 1 (1952): 18–49.

"Quanguo zaolinyundong quanmian zhankai, Sichuan Guangxi chao'e wancheng quannian zaolinjihua" 全国造林运动全面展开四川广西超额完成全年造林计划 [Nationwide afforestation campaign fully launched, Sichuan and Guangxi exceed annual afforestation plan targets]. *Renmin ribao* 人民日报, April 28, 1959.

"Ruhe pingjia zhifubao *Mayisenlin* 3nian zhong zhenshu 1yi ke, weishenme neng xiyin zhemeduoren yiqi zhongshu?" 如何评价支付宝《蚂蚁森林》3年种真树1亿棵, 为什么能吸引这么多人一起种树? [How do you evaluate that Alipay *Ant Forest* has planted 100 million real trees in three years, why can it attract so many people to plant trees together?]. Zhihu 知乎, April 22, 2019. https://www.zhihu.com/question/321188877.

Sha Qing 沙青. "Beijing shiqupingheng" 北京失去平衡 [Beijing out of balance]. *Xinhuawenzhai* 新华文摘 12, no. 6 (1986): 123–36.

"Shamo lüzhou Suji tan" 沙漠绿洲苏吉滩 [Desert oasis in Suji]. *Neimenggu ribao* 内蒙古日报, December 8, 1974.

"Shamo xinge" 沙漠新歌 [New songs from the desert]. *Neimenggu ribao* 内蒙古日报, October 15, 1974.

"Shanxi baiwan qingshaonian kaizhan tujiyundong yigeyuenei lühua huanghe-fenhe" 山西百万青少年开展突击运动一月内绿化黄河汾河 [Millions of teenagers in Shanxi launch assault campaign to afforest the Fenhe River of the Yellow River within one month]. *Renmin ribao* 人民日报, March 5, 1958.

Sheban linchang fang guangman 社办林场放光芒 [The people's commune forest farm shines brightly]. Beijing: Zhonguo linye chubanshe, 1959.

Shennikov, A. P. *Zhiwu shengtaixue* 植物生态学 [Plant ecology]. Translated by Wang Wen. Shanghai: Xinnongchubanshe, 1953.

"Shouming zuge de He Yingqin jiangjun" 受命组阁的何应钦将军 [General He Yingqin appointed to form a cabinet]. *Xinwen tiandi* 新闻天地, no. 62 (1949): 1.

"Shuilibu Fu Zuoyi buzhangdeng kancha huanghe xushuikuzhi wanbi" 水利部傅作义部长等勘查黄河蓄水库址完毕 [Minister Fu Zuoyi of the Ministry of Water Resources has completed a survey of the Yellow River reservoir site]. *Renmin ribao* 人民日报, July 25, 1950.

"Sidalin pailaideren zenyangzai zhongguode heliushang gongzuozhe" 斯大林派来的人怎样在中国的河流上工作着 [How Stalin's experts work on China's rivers]. *Renmin ribao* 人民日报, November 25, 1951.

State Council Information Office of the People's Republic of China. "The History and Development of the Xinjiang Production and Construction Corps." Guowuyuan xinwen bangongshi 国务院新闻办公室, October 5, 2014. https://english.www.gov.cn/archive/white_paper/2014/10/05/content_281474992384669.htm.

Sun Zhongshan [Sun Yat-sen] 孙中山. "Jianguo fanglüe" 建国方略 [Plan for building the nation]. In *Sunzhongshan quanji* 孙中山全集 [Complete works of Sun Yat-sen], vol. 6, edited by Guangdongsheng shehuikexueyuan 广东省社会科学院, Zhongguo shekeyuan 中国社科院, and Zhongshan daxue 中山大学, 157–493. Beijing: Zhonghuashuju, 1985.

———. "Sanminzhuyi minshengzhuyi" 三民主义民生主义 [The three principles of the people, the principle of people's livelihood]. In *Sunzhongshan quanji* 孙中山全集 [Complete works of Sun Yat-sen], vol. 9, edited by Guangdongsheng shehuikexueyuan 广东省社会科学院, Zhongguo shekeyuan 中国社科院, and Zhongshan daxue 中山大学, 183–426. Beijing: Zhonghuashuju, 1986.

———. "Zai shanghai Zhongguo shehuidangde yanshuo" 在上海中国社会党的演说 [Speech at the Chinese Socialist Party in Shanghai]. In *Sunzhongshan quanji* 孙中山全集 [Complete works of Sun Yat-sen], vol. 2, edited by Guangdongsheng shehuikexueyuan 广东省社会科学院, Zhongguo shekeyuan 中国社科院, and Zhongshan daxue 中山大学, 343. Beijing: Zhonghuashuju, 1982.

———. "Zhongguo guomindang xuanyan" 中国国民党宣言 [Chinese Nationalist Party declaration]. In *Sunzhongshan quanji* 孙中山全集 [Complete works of Sun Yat-sen], vol. 7, edited by Guangdongsheng shehuikexueyuan 广东省社会科学

院, Zhongguo shekeyuan 中国社科院, and Zhongshan daxue 中山大学, 1–4. Beijing: Zhonghuashuju, 1985.

———. "Zhongguo zhi tielu jihua yu minshengzhuyi" 中国之铁路计划与民生主义 [China's railway plan and principles of people's livelihood]. In *Sunzhongshan quanji* 孙中山全集 [Complete works of Sun Yat-sen], vol. 2, edited by Guangdongsheng shehuikexueyuan 广东省社会科学院, Zhongguo shekeyuan 中国社科院, and Zhongshan daxue 中山大学, 487–593. Beijing: Zhonghuashuju, 1982.

Tao Bennai 陶本艿. "Qihougongsi de gushi" 气候公司的故事 [The story of the Climate Company]. In *Zhongguo kehuanxiaoshuo daquan* 中国科幻小说大全 [The complete collection of Chinese science fiction novels], edited by Rao Zhonghua 饶忠华, 99–100. Beijing: Haiyang chubanshe, 1982.

T'sen Shen [Cen Shen 岑参]. "A Song of the Running Horse River: Presented on Saying Farewell to the Army Going on Campaign to the West." In *The Columbia Anthology of Traditional Chinese Literature*, edited by Victor Mair, translated by Daniel Bryant, 101. New York: Columbia University Press, 2000.

Wang Junling 王俊岭. "5.5yishuang jiao 'zhong'chule zhepian 'mayisenlin'" 5.5亿双脚"种"出了这片"蚂蚁森林" [550 million pairs of feet "planted" this "ant forest"]. *Renminwang* 人民网, June 29, 2020. http://paper.people.com.cn/rmrbhwb/html/2020-06/29/content_1994419.htm.

Wang Meng 王蒙. "Zanmei lüye" 赞美绿叶 [Praising green leaves]. *Lüye* 绿叶1, no. 1 (1992): 7.

Wang Pixian 王丕显. "Duiyu zhishujie yingxing zhengdun zhi yijian" 对于植树节应行整顿之意见 [Suggestions to rectify Arbor Day]. *Jiangsu shiye yuezhi* 江苏实业月志, no. 4 (1926): 118–23.

Wang Taiyue 王太岳. "Tongshanyin" 铜山吟 [Lament for copper-bearing hills]. In *Qingshiduo* 清诗铎 [Qing bell of poesy], edited by Zhang Yingchang 张应昌, 926–28. Beijing: Zhonghuashuju, 1960.

Wang Xiaolin 王晓琳. "Lishi yu dierdai senlin" 林业与第二代森林 [History and second-generation forests]. *Linye wenxue* 林业文学 3, no. 3 (June 1989): 11–17.

"Wanlishamo pilüzhuang" 万里沙漠披绿装 [Covering thousands of miles of desert with green cover]. *Neimenggu ribao* 内蒙古日报, October 15, 1974.

"Woguodiyibu linye dianyingpian" 我国第一部林业电影片 [China's first forestry film]. In *Zhongguo linye zhizui* 中国林业之最 [The most important issues of Chinese forestry], edited by Tang Shengxiang 汤升亨 and Tang Yunli 唐云丽, 143–44. Beijing: Zhongguo linye chubanshe, 1993.

Wu Baosan 吴宝三 and Cao Feng 曹锋. *Ma Yongshun zhuan* 马永顺传 [Biography of Ma Yongshun]. Harbin: Heilongjiang renmin chubanshe, 1999.

Wu Jing 吴晶 and Yang Yijun 杨依军. "Xi Jinping zai canjia shoudu yiwuzhishu huodong shi qiangdiao: Laogu shuli lüshuiqingshan jiushi jinshanyinshan linian dazao qingshanchangzai lüshuichangliu kongqichangxin meilizhongguo" 习近平在参加首都义务植树活动时强调：牢固树立绿水青山就是金山银山理念打造青山常在绿水长流空气常新美丽中国 [During the capital's voluntary tree-planting activity, Xi Jinping emphasizes that we must firmly establish the concept that lucid waters and lush mountains are invaluable assets and strive to create a beautiful China with evergreen mountains, perennially flowing waters, and ever-fresh air]. *Xinhuanet* 新华网, April 3, 2020. http://www.xinhuanet.com/politics/leaders/2020-04/03/c_1125811438.htm.

Wu Shuchen 吴庶晨. "Zhishushi yu zhishujie" 植树式与植树节 [Tree-planting ceremony and Arbor Day]. *Xinwen bao* 新闻报, March 12, 1929.

Xiao Jianheng 肖建亨. "Milin hu zong" 密林虎踪 [Tiger traces in the forest]. *Shaonian kexue* 少年科学 2, no. 4 (1977): 13–24.

"Xishan zhishujie zhi shenghui" 西山植树节之盛会 [The grand celebration of Arbor Day at Xishan]. *Yishibao (Tianjinban)* 益世报 (天津版), April 7, 1916.

Xu Gang 徐刚. *Da senlin* 大森林 [Big forest]. Beijing: Beijing shiyue wenyi chubanshe, 2017.

———. *Diqiu zhuan* 地球传 [Biography of the earth]. Beijing: Zuojia chubanshe, 2009.

———. *Famuzhe, xinglai!* 伐木者,醒来! [Woodcutter, wake up!]. Changchun: Jilin renmin chubanshe, 1997.

———. *Guonan* 国难 [National disaster]. Beijing: Zuojia chubahsen, 2005.

———. *Lüse xuanyuan* 绿色宣言 [Green proclamation]. Changchun: Shidai wenyi chubanshe, 1997.

———. "Ziran biji" 自然笔记 [Notebook of nature]. *Renmin wenxue* 人民文学 74, no. 1 (2022): 4–21.

Xun Min 勋民. "Zhishu yundong" 植树运动 [Tree-planting movement]. *Xinwen bao* 新闻报, March 12, 1928.

Yang Fu 仰夫. "Zhishu yu chuxu" 植树与储蓄 [Planting and saving trees]. *Xinwen bao* 新闻报, March 12, 1928.

Yang Xunan 杨叙男 and Tao Jinsheng 陶锦生. "Milin shaobing" 密林哨兵 [Scout in the jungle]. *Shaonian kexue* 少年科学 4, no. 4 (1979): 12–22.

Yang Zhenya 杨振亚 and Deng Pu 邓普, dir. *Junken zhange* 军垦战歌 [The army's reclamation and battle song]. Beijing: August First Film Studio, 1965.

Yano Munemoto 矢野宗幹. *Kunchong shengtaixue* 昆虫生态学 [Insect ecology]. Translated by Xue Deyu 薛德焴. Beijing: Shangwu yinshuguan, 1935.

You Xueqing 游雪晴. "Sanbei fanghulin dangzhu fengchuiwumai? Zhuanjia: Shu bubeizhe 'guo'" 三北防护林挡住风吹雾霾？专家：树不背这"锅" [Does the Three-North Shelter belt block the wind and smog? Expert: trees should be the "scapegoat"]. Huanqiuwang 环球网, January 18, 2017. https://tech.huanqiu.com/article/9CaKrnJZQjl.

"Youganyu zhishujie" 有感于植树节 [Reflections on Arbor Day]. *Shishi xinbao wankan* 时事新报晚刊, March 14, 1948.

Yuan Zhen 元稹. "Duixiudiqingzhongshupan" 对修堤请种树判 [Judgment on the request to plant trees for dike repair]. In *Zhongguo gudai linye wenxian xuandu* 中国古代林业文献选读 [Selected readings from ancient Chinese forestry literature], edited by Zhang Lianwei 张连伟, Li Fei 李飞, and Zhou Jingyong 周景勇, 193–94. Beijing: Beijing yanshan chubanshe, 2015.

"Yuesheng fangshui yu zaolin" 粤省防水与造林 [Flood prevention and afforestation in Guangdong Province]. *Shenbao* 申报, May 6, 1949.

Yun Feng 云峰 and Zhuletemu 朱勒特木. "Chuntian shuyu ganyu douzheng de ren" 春天属于敢斗争的人 [Spring belongs to those who dare to fight bravely]. *Renmin dianying* 人民电影 1, no. 2 (1976): 25–26.

Zai linhai shenchu 在林海深处 [In the depths of the forests]. Beijing: Zhonguo linye chubanshe, 1960.

"Zaolinzhishu dianli zhongshuyaoren duoqinwang canjia shenbuzhang juchu sanzhongyiyi" 造林植树典礼中枢要人多亲往参加沈部长举出三种意义 [Afforestation and tree-planting ceremony: many important figures from the central government attended in person; Minister Shen elaborates on three significances]. *Zhongyang ribao (Chongqing)* 中央日报(重庆), March 13, 1942.

"Zengzhishumu baohushanlin" 增植树木保护山林 [Plant more trees and protect mountains and forests]. *Hongse Zhonghua* 红色中华, no. 150 (1934): 3.

Zhang Ge 张革. *Milin li de zhandou* 密林里的战斗 [Fighting in the jungle]. Shanghai: Shaonianertong chubanshe, 1959.

Zhang Hanying 张含英. "Xin Huanghe de guangmingqiantu: Shuilizhuanjia Zhang Hanying zai shuili lianxihuiyishang de baogaozhaiyao" 新黄河的光明前途: 水利专家张含英在水利联席会议上的报告摘要 [The bright future of the new Yellow River: Zhang Hanying's report at the joint meeting of water conservancy experts]. *Renmin ribao* 人民日报, November 15, 1949.

———. *Zhiheluncong* 治河论丛 [Collections of essays on river management]. Beijing: Shangwu yinshuguan, 1936.

Zhang Ting 张珽 and Dong Shuangqiu 董爽秋. *Zhiwu shengtaixue* 植物生态学 [Plant ecology]. Guangzhou: Guangzhou weixing yinshuachang, 1930.

Zhang Ziping 张资平. *Zhishujie* 植树节 [Arbor Day]. Shanghai: Xinyuzhou shudian, 1928.

Zhao Jiting 赵骥亭. "Lun zhishujie zhi youming wushi" 论植树节之有名无实 [On Arbor Day being in name only, not in reality]. *Yingqiu* 嘤求, no. 4 (1928): 6–8.

"Zhengshi: Xuzhishuxueshu" 政事: 续植树学塾 [Political affairs: continuing the tree-planting school]. *Wanguo gongbao* 万国公报, no. 504 (1878): 3–4.

"Zhifubaode *Mayisenlin* shizhende dui shengtai chansheng le bangzhume, haishi zhishi biaomian gongfu?" 支付宝的《蚂蚁森林》是真的对生态产生了帮助么，还是只是做表面功夫? [Does Alipay's *Ant Forest* really help the ecology, or is it just a superficial effort?]. Zhihu 知乎, March 6, 2018. https://www.zhihu.com/question/268277647.

"Zhishu" 植树 [Tree planting]. *Guowen zhoubao* 国闻周报11, no. 12 (1934): 1.
"Zhishu biaoyu" 植树标语 [Tree-planting slogan]. Special issue, *Xinwen bao* 新闻报, March 12, 1928.
"Zhishu xueshu" 植树学塾 [Tree-planting school]. *Wanguo gongbao* 万国公报, no. 502 (1878): 2–3.
"Zhishu yi jinian zhongshanxiansheng" 植树以纪念中山先生 [Plant a tree in memory of Sun Yat-sen]. *Shibao* 时报, March 12, 1928.
"Zhishu youguan" 植树有感 [Reflections on tree planting]. *Zhongyang ribao* 中央日报, March 14, 1946.
"Zhishujie dianlinggaiqi" 植树节电令改期 [Arbor Day telegraphic order for rescheduling]. *Shengjingshibao* 盛京时报, March 22, 1929.
"Zhishujie jinianhui" 植树节纪念会 [Arbor Day commemoration party]. *Lianhebao* 联合报, March 12, 1952.
"Zhishujiepeidujuxing zhishudianli" 植树节陪都举行植树典礼 [The capital holds a tree-planting ceremony on Arbor Day]. *Xijing shibao* 西京时报, March 13, 1942.
"Zhishuzaolin, lühuazuguo" 植树造林,绿化祖国 [Planting trees, greening the motherland]. *Renminhuabao* 人民画报 27, no. 9 (September 1976): 15–17.
Zhong Ming 中铭. "Zhenwang jiangshi zhuidao dahui zhishudianli" 阵亡将士追悼大会植树典礼 [Tree-planting ceremony at memorial meeting for fallen soldiers]. *Shishi yuebao* 时事月报 4, no. 4 (1931): 13.
Zhongguo huanjingnianjian bianji weiyuanhui 中国环境年鉴编辑委员会, ed. *Zhongguo huanjingnianjian 1992* 中国环境年鉴1992 [China environment yearbook 1992]. Beijing: Zhongguo huanjingkexue chubanshe, 1992.
Zhongguo linye [Chinese forestry] 1, no. 1 (June 1950).
"Zhongguode mayijinfu zhanshi le ruhe tongguo shuzi yingxiangli yingdui qihoubianhua" 中国的蚂蚁金服展示了如何通过数字影响力应对气候变化 [China's Ant Financial demonstrates how to tackle climate change through digital influence]. Lianheguohuanjingguihuashu 联合国环境规划署, November 3, 2017. https://www.unenvironment.org/zh-hans/xinwenyuziyuan/gushi/zhongguodemayijinfuzhanshileruhetongguoshuziyingxiangliyingduiqihoubianhua.
Zhonghuarenmingongheguo guowuyuan 中华人民共和国国务院. "1956 nian dao 1967 nian quanguonongye fazhan gangyao (cao'an)" 1956年到1967年全国农业发展纲要 (草案) [National agricultural development outline from 1956 to 1967 (draft)]. Zhongguo zhengfuwang 中国政府网, February 6, 1956. https://www.gov.cn/gongbao/shuju/1956/gwyb195605.pdf.
Zhonghuarenmingongheguo linyebu 中华人民共和国林业部, ed. *Lühua zuguo de gaochao* 绿化祖国的高潮 [The climax of greening the motherland]. Vol. 1. Beijing: Beijing linye chubanshe, 1958.
"Zhonghuasenlinhui zhi guohui geyiyuanshu" 中华森林会致国会各议员书 [Letter from the Chinese Forestry Association to the members of the national congress]. *Shenbao* 申报, April 8, 1917.

Zhou Enlai and Jawaharlal Nehru. "Minutes of the First Meeting between Premier Zhou Enlai and Nehru." October 19, 1954, PRC FMA 204-00007-03, 14–26. Wilson Center Digital Archive, March 23, 2015. https://digitalarchive.wilsoncenter.org/document/121746.

Zhou Hui'an 周晦龠. *Zhengfu ziran* 征服自然 [Conquering nature]. Beiping: Xiehua chubanshe, 1932.

———. *Zhishu jie* 植树节 [Arbor Day]. Beiping: Xiehua chubanshe, 1932.

Zhou Libo 周立波. *Baofengzhouyu* 暴风骤雨 [Hurricane]. Shijiazhuang: Huashan wenyi chubanshe, 1995.

Zhou Qunfeng 周群峰. "Qingdao 'zhongshushizhang' Zhang Xinqi tuixiuhoubeicha, chuxihuodongshi yipianxusheng" 青岛"种树市长"张新起退休三年后被查，出席活动时市民一片嘘声 [Qingdao "tree-planting mayor" Zhang Xinqi was investigated three years after retiring, and the citizens booed when he attended the event]. *Zhongguo xinwenzhoukan* 中国新闻周刊, March 12, 2021. https://www.chinanews.com/gn/2021/03-12/9430456.shtml.

Zhu Changlong 褚长龙. "Kanyingpian *Junken zhange* de ganxiang" 看影片《军垦战歌》的感想 [Reflections on watching *The army's reclamation and battle song*]. *Dianyingyishu* 电影艺术 9, no. 6 (1965): 63.

Zhu Kezhen 竺可桢. "Gaizao shamo shi womende lishirenwu" 改造沙漠是我们的历史任务 [Transforming the desert is our historical mission]. *Renmin ribao* 人民日报, March 2, 1959.

———. "Xiangshamo jinjun" 向沙漠进军 [Marching into the desert]. *Renmin ribao* 人民日报, February 9, 1961.

Zhu Wenshu 朱文叔 and Lü Boyou 吕伯攸, eds. *Xiaoxueguoyukeben: Chuji* 小学国语课本：初级 [Primary school Chinese textbook: elementary level]. Shanghai: Zhonghua shuju, 1936.

Zhu Wenshun 朱文顺 and Liu Zhongming 刘中明, dir. *Shamo de chuntian* 沙漠的春天 [Spring in the desert]. Changchun: Changchun Studio, 1975.

"Zhuajinshiji tuji qiuji lühuazaolin" 抓紧时机突击秋季绿化造林 [Seize the opportunity to carry out fall greening and afforestation]. *Xinjiang ribao* 新疆日报, October 30, 1959.

"Zizhongqu chao'e wancheng jinnian zhisha renwu" 自治区超额完成今年治沙任务 [The autonomous region overfulfilled this year's desert control tasks]. *Xinjiang ribao* 新疆日报, October 30, 1959.

"Zongli shishi wuzhounian zhishu jinian: Zhishu qingxing" 总理逝世五周年植树纪念：植树情形 [Tree-planting commemoration of the fifth anniversary of the death of the prime minister: tree-planting situation]. *Nanxun tielu yuebao* 南浔铁路月报 8, no. 3 (1930): 1.

Secondary Sources

Anderson, Benedict. *Imagined Communities: Reflections on the Origin and Spread of Nationalism*. New York, NY: Verso Books, [1983] 2006.

Auricchio, Laura, Elizabeth Heckendorn Cook, and Giulia Pacini, eds. *Invaluable Trees: Cultures of Nature, 1660–1830*. Oxford: Voltaire Foundation, 2012.

Barton, Gregory Allen. *Empire Forestry and the Origins of Environmentalism*. New York, NY: Cambridge University Press, 2002.

Bello, David. *Across Forest, Steppe, and Mountain: Environment, Identity, and Empire in Qing China's Borderlands*. New York, NY: Cambridge University Press, 2015.

———. "Splintered Habitats: The Fragmentation of Ecotone Northern China's Imperial Woodland Complexes." In *The Cultivated Forest: People and Woodlands in Asian History*, edited by Ian M. Miller, Bradley Camp Davis, Brian Lander, and John S. Lee, 95–119. Seattle: University of Washington Press, 2022.

Bergère, Marie-Claire. *Sun Yat-sen*. Translated by Janet Lloyd. Stanford, CA: Stanford University Press, 1998.

Bernards, Brian. *Writing the South Seas: Imagining the Nanyang in Chinese and Southeast Asian Postcolonial Literature*. Seattle: University of Washington Press, 2015.

Biggs, David. "Managing a Rebel Landscape: Conservation, Pioneers, and the Revolutionary Past in the U Minh Forest, Vietnam." *Environmental History* 10, no. 3 (2005): 448–76.

Brady, Lisa M. "Sowing War, Reaping Peace: United Nations Resource Development Programs in the Republic of Korea, 1950–1953." *Journal of Asian Studies* 77, no. 2 (2018): 351–63.

Braester, Yomi. *Painting the City Red: Chinese Cinema and the Urban Contract*. Durham, NC: Duke University Press, 2010.

Brain, Stephen. *Song of the Forest: Russian Forestry and Stalinist Environmentalism, 1905–1953*. Pittsburgh, PA: University of Pittsburgh Press, 2011.

———. "Stalin's Environmentalism." *Russian Review* 69, no. 1 (2010): 93–118.

Brain, Stephen, and Viktor Pál, eds. *Environmentalism under Authoritarian Regimes: Myth, Propaganda, Reality*. New York, NY: Routledge, 2019.

Brophy, David. *Uyghur Nation: Reform and Revolution on the Russia-China Frontier*. Cambridge, MA: Harvard University Press, 2016.

Brown, Karen. "'Trees, Forests and Communities': Some Historiographical Approaches to Environmental History on Africa." *Area* 35, no. 4 (2003): 343–56.

Brown, Kate. *Plutopia: Nuclear Families, Atomic Cities, and the Great Soviet and American Plutonium Disasters*. Oxford: Oxford University Press, 2013.

Brown, Tristan. *Laws of the Land: Fengshui and the State in Qing Dynasty China*. Princeton, NJ: Princeton University Press, 2023.

Brunner, Elizabeth. *Environmental Activism, Social Media, and Protest in China: Becoming Activists over Wild Public Networks*. New York, NY: Lexington Books, 2019.

Bruun, Ole. *Fengshui in China: Geomantic Divination between State Orthodoxy and Popular Religion*. Honolulu: University of Hawai'i Press, 2003.

Butler, Judith. *Precarious Life: The Powers of Mourning and Violence*. New York, NY: Verso Books, 2006.

Byrnes, Corey. *Fixing Landscape: A Techno-Poetic History of China's Three Gorges*. New York, NY: Columbia University Press, 2019.

Cao, Shixiong. "Why Large-Scale Afforestation Efforts in China Have Failed to Solve the Desertification Problem." *Environmental Science and Technology* 42, no. 6 (March 2008): 1826–31.

Chamovitz, Daniel. *What a Plant Knows: A Field Guide to the Senses* (New York, NY: Farrar, Strauss and Giroux, 2012).

Chan, Cheow Thia. *Malaysian Crossings: Place and Language in the Worlding of Modern Chinese Literature*. New York, NY: Columbia University Press, 2022.

Chang, Alenda. *Playing Nature: Ecology in Video Games*. Minneapolis: University of Minnesota Press, 2019.

Chang, Chia-ju, ed. *Chinese Environmental Humanities: Practices of Environing at the Margins*. New York, NY: Springer, 2019.

Chen, Geoffrey, and Charles Lees. "The New, Green, Urbanization in China: Between Authoritarian Environmentalism and Decentralization." *Chinese Political Science Review* 3, no. 2 (2018): 212–31.

Chen, Tina Mai. "Propagating the Propaganda Film: The Meaning of Film in CCP Writings, 1949–1965." *Modern Chinese Literature and Culture* 15, no. 2 (Fall 2003): 154–93.

Chen, Xiaomei. *Occidentalism: A Theory of Counter-Discourse in Post-Mao China*. Oxford: Oxford University Press, 1995.

Chen Xiaoming 陈晓明. *Biaoyi de jiaolü* 表意的焦虑 [The anxiety of expression]. Beijing: Zhongyang bianyi chubanshe, 2004.

Chen Yunqian 陈蕴茜. "Zhishujie yu sunzhongshan chongbai" 植树节与孙中山崇拜 [Arbor Day and the worship of Sun Yat-sen]. *Nanjing daxue xuebao* 南京大学学报 13, no. 5 (May 2006): 76–90.

Cheng Guangwei 程光炜. *Chongfan bashi niandai* 重返八十年代 [Revisiting the 1980s]. Beijing: Beijingdaxue chubanshe, 2009.

Chi Xiang 池翔. "Chongsubianjiang: Yalujiang youan de linmu caifa senlinjiaosheyubianjiangzhixu" 重塑边疆:鸭绿江右岸的林木采伐森林交涉与边疆秩序 [Reshaping the frontier: timber harvesting, forest negotiations, and frontier order on the right bank of the Yalu River]. *Chongqingdaxuexuebao* 重庆大学学报 5 (2020): 265–78.

Clinton, Maggie. *Revolutionary Nativism: Fascism and Culture in China, 1925–1937*. Durham, NC: Duke University Press, 2017.

Cohen, Paul. *History in Three Keys: The Boxers as Event, Experience, and Myth*. New York, NY: Columbia University Press, 1997.

Cohen, Shaul E. *Planting Nature: Trees and the Manipulation of Environmental Stewardship in America*. Berkeley: University of California Press, 2004.

Cronon, William. "A Place for Stories: Nature, History, and Narrative." *Journal of American History* 78, no. 4 (1992): 1347–76.

———. "The Trouble with Wilderness; or, Getting Back to the Wrong Nature." *Environmental History* 1, no. 1 (1996): 7–28.

Crosby, Alfred. *Ecological Imperialism: The Biological Expansion of Europe, 900–1900*. New York, NY: Cambridge University Press, 2004.

Cummins, Elizabeth. *Understanding Ursula K. Le Guin*. Columbia: University of South Carolina Press, 1990.

Dauvergne, Peter. *Environmentalism of the Rich*. Cambridge, MA: MIT Press, 2016.

Davis, Diana. *Resurrecting the Granary of Rome: Environmental History and French Colonial Expansion in North Africa*. Athens: Ohio University Press, 2007.

Dikötter, Frank. *Mao's Great Famine: The History of China's Most Devastating Catastrophe, 1958–1962*. New York, NY: Bloomsbury, 2010.

Dillon, Michael. *Xinjiang: China's Muslim Far Northwest*. New York, NY: Routledge, 2004.

Ding, Iza. *The Performative State: Public Scrutiny and Environmental Governance in China*. Ithaca, NY: Cornell University Press, 2022.

Eaton, Sarah, and Genia Kostka. "Authoritarian Environmentalism Undermined? Local Leaders' Time Horizons and Environmental Policy Implementation in China." *China Quarterly* 218 (June 2014): 359–80.

Economy, Elizabeth. *The River Runs Black: The Environmental Challenge to China's Future*. Ithaca, NY: Cornell University Press, 2011.

Egleston, Nathaniel Hillyer. *Arbor Day: Its History and Observance*. Report (US Department of Agriculture) 56. Washington, DC: US Government Printing Office, 1896.

Elvin, Mark. *The Retreat of the Elephants: An Environmental History of China*. New Haven, CT: Yale University Press, 2008.

Engerman, David. *The Price of Aid: The Economic Cold War in India*. Cambridge, MA: Harvard University Press, 2018.

Esherick, Joseph W. "How the Qing Became China." In *Empire to Nation: Historical Perspectives on the Making of the Modern World*, edited by Joseph W. Esherick, Hasan Kayali, and Eric Van Young, 229–59. New York, NY: Lexington Books, 2009.

Estok, Simon C. *The Ecophobia Hypothesis*. New York, NY: Routledge, 2018.

Esty, Daniel C., and Andrew Winston. *Green to Gold: How Smart Companies Use Environmental Strategy to Innovate, Create Value, and Build Competitive Advantage*. New Haven, CT: Yale University Press, 2006.

Fedman, David. *Seeds of Control: Japan's Empire of Forestry in Colonial Korea*. Seattle: University of Washington Press, 2020.

Feng, Qi, Hua Ma, Xuemei Jiang, Xin Wang, and Shixiong Cao. "What Has Caused Desertification in China?" *Scientific Reports* 5 (2015): 15998. https://doi.org/10.1038/srep15998.

Fredriksson, Per G., and Jim R. Wollscheid. "Democratic Institutions versus Autocratic Regimes: The Case of Environmental Policy." *Public Choice* 130, no. 3 (2007): 381–93.

Freud, Sigmund. "Mourning and Melancholia." *Journal of Nervous and Mental Disease* 56, no. 5 (1922): 543–45.

Frye, Northrop. *A Natural Perspective*. New York, NY: Columbia University Press, 1965.

Gallicano, Tiffany Derville. "A Critical Analysis of Greenwashing Claims." *Public Relations Journal* 5, no. 3 (2011): 1–21.

Gilley, Bruce. "Authoritarian Environmentalism and China's Response to Climate Change." *Environmental Politics* 21, no. 2 (2012): 287–307.

Goldstein, Joshua. *Remains of the Everyday: A Century of Recycling in Beijing*. Berkeley: University of California Press, 2021.

Golley, Frank B. *A History of the Ecosystem Concept in Ecology: More than the Sum of the Parts*. New Haven, CT: Yale University Press, 1993.

Groppe, Alison M. *Sinophone Malaysian Literature: Not Made in China*. Amherst, MA: Cambria Press, 2013.

Gross, Miriam. *Farewell to the God of Plague: Chairman Mao's Campaign to Deworm China*. Berkeley: University of California Press, 2016.

Grove, Richard. *Green Imperialism: Colonial Expansion, Tropical Island Edens and the Origins of Environmentalism, 1600–1860*. New York, NY: Cambridge University Press, 1995.

Guha, Ramachandra. "Forestry in British and Post-British India: A Historical Analysis." *Economic and Political Weekly* 18, no. 44 (1983): 1882–96.

———. *The Unquiet Woods: Ecological Change and Peasant Resistance in the Himalaya*. Berkeley: University of California Press, 1990.

Guo, Li. "From the Margins to the Hearth." *Modern Chinese Literature and Culture* 31, no. 2 (2019): 162–206.

Han, Heejin. "Authoritarian Environmentalism under Democracy: Korea's River Restoration Project." *Environmental Politics* 24, no. 5 (2015): 810–29.

Hannigan, Robert E. *The Great War and American Foreign Policy, 1914–24*. Philadelphia: University of Pennsylvania Press, 2017.

Haraway, Donna. "Anthropocene, Capitalocene, Plantationocene, Chthulucene: Making Kin." *Environmental Humanities* 6, no. 1 (2015): 159–65.

Harper, Davis. "Starbucks Falls Short on Environmental Commitments: Reports Suggest That Caffeine Lovers' Iconic Mermaid May Not Be So Green After All." *Sierra*, November 21, 2017. https://www.sierraclub.org/sierra/starbucks-falls-short-environmental-commitments.

Harrell, Stevan. *An Ecological History of Modern China*. Seattle: University of Washington Press, 2023.

Harrison, Henrietta. *The Making of the Republican Citizen: Political Ceremonies and Symbols in China 1911–1929*. New York, NY: Oxford University Press, 2000.

Harrison, Robert Pogue. *Forests: The Shadow of Civilization*. Chicago: University of Chicago Press, 1992.

He Guimei 贺桂梅. *"Xinqimeng" zhishidangan: Bashiniandai Zhongguo wenhua yanjiu* "新启蒙"知识档案: 八十年代中国文化研究 ["New Enlightenment" knowledge archives: Chinese cultural studies in the 1980s]. Beijing: Beijing daxue chubanshe, 2010.

He, Qiliang. *The People's West Lake: Propaganda, Agency, and Nature in Mao's China, 1949–1976*. Honolulu: University of Hawai'i Press, 2023.

Hehn, Paul N. "Serbia, Croatia and Germany 1941–1945: Civil War and Revolution in the Balkans." *Canadian Slavonic Papers* 13, no. 4 (1971): 344–73.

Heise, Ursula. *Imagining Extinction: The Cultural Meanings of Endangered Species*. Chicago: University of Chicago Press, 2016.

———. *Sense of Place and Sense of Planet: The Environmental Imagination of the Global*. New York, NY: Oxford University Press, 2008.

Heske, Franz. *German Forestry*. New Haven, CT: Yale University Press, 1938.

Hinton, William. *Fanshen: A Documentary of Revolution in a Chinese Village*. New York, NY: Monthly Review Press, 1967.

Hong Zicheng 洪子诚. *Zhongguo wenxue 1949–1989* 中国文学1949–1989 [Chinese literature 1949–1989]. Beijing: Beijing chubanshe, 2020.

Honig, Emily, and Xiaojian Zhao. *Across the Great Divide: The Sent-Down Youth Movement in Mao's China, 1968–1980*. New York, NY: Cambridge University Press, 2019.

Horkheimer, Max, and Theodor W. Adorno. *Dialect of Enlightenment: Philosophical Fragments*. Translated by Edmund Jephcott. Stanford, CA: Stanford University Press, 2002.

Hou Li. *Building for Oil: Daqing and the Formation of the Chinese Socialist State*. Cambridge, MA: Harvard University Asia Center, 2018.

Huang Yanzhong. *Toxic Politics: China's Environmental Health Crisis and Its Challenge to the Chinese State*. New York, NY: Cambridge University Press, 2020.

Huang Zhenglin 黄正林. *Shaan Gan Ning bianqu shehuijingjishi (1937–1945)* 陕甘宁

边区社会经济史 (1937–1945) [Social and economic history of the Shaanxi-Gansu-Ningxia border region (1937–1945)]. Beijing: Renmin chubanshe, 2006.

Huang Ziping 黄子平. "Guanyu 'weixiandaipai' jiqi piping" 关于"伪现代派"及其批评 [On "pseudomodernism" and its criticism]. *Beijing wenxue* 北京文学 2 (1988): 4.

Huntington, Samuel P. *Political Order in Changing Societies*. New Haven, CT: Yale University Press, 1968.

Hutchinson, G. Evelyn. "Circular Causal Systems in Ecology." *Annals of the New York Academy of Sciences* 50 (1948): 221–46.

Iheka, Cajetan. *African Ecomedia: Network Forms, Planetary Politics*. Durham, NC: Duke University Press, 2021.

———. *Naturalizing Africa: Ecological Violence, Agency, and Postcolonial Resistance in African Literature*. New York, NY: Cambridge University Press, 2018.

Ivanhoe, Philip J. "Early Confucianism and Environmental Ethics." In *Confucianism and Ecology: The Interrelation of Heaven, Earth, and Humans*, edited by Mary Tucker and John Berthrong, 59–76. Cambridge, MA: Harvard University Press, 1998.

Jagoda, Patrick. *Experimental Games: Critique, Play, and Design in the Age of Gamification*. Chicago: University of Chicago Press, 2020.

Jiang Xin 江心, Wang Xiqun 王希群, Guo Baoxiang 郭保香, Hu Yong 胡涌, and Liu Changhai 刘长海. "Shaan Gan Ning bianqu linyefazhanshiyanjiu 1937–1950" 陕甘宁边区林业发展史研究1937–1950 [Research on the history of forestry development in the Shaanxi-Gansu-Ningxia border region 1937–1950]. *Beijing linye daxue xuebao* 北京林业大学学报11, no. 1 (2012): 1–24.

Khan, Almaz. "Who Are the Mongols? State, Ethnicity, and the Politics of Representation in the People's Republic of China." In *Negotiating Ethnicities in China and Taiwan*, edited by Melissa Brown, 125–59. Berkeley: University of California Press, 1996.

Kingsland, Sharon E. *The Evolution of American Ecology, 1890–2000*. Baltimore: Johns Hopkins University Press, 2005.

Kinzley, Judd. *Natural Resources and the New Frontier: Constructing Modern China's Borderlands*. Chicago: University of Chicago Press, 2018.

Kirby, William. *Empires of Ideas: Creating the Modern University from Germany to America to China*. Cambridge, MA: Harvard University Press, 2022.

———. *Germany and Republican China*. Stanford, CA: Stanford University Press, 1984.

Klinek, Eric. "Arbor Day: Educating Environmental Citizens." *Pennsylvania Legacies* 10, no. 1 (2010): 3–5.

Klingle, Matthew. *Emerald City: An Environmental History of Seattle*. New Haven, CT: Yale University Press, 2007.

Knott, John. *Imagining the Forest: Narratives of Michigan and the Upper Midwest.* Ann Arbor: University of Michigan Press, 2012.

Kuang Xinnian 旷新年. *Xiezai dangdai wenxue bianshang* 写在当代文学边上 [Written on the edge of contemporary literature]. Shanghai: Shanghai jiaoyu chubanshe, 2005.

Lander, Brian. "Deforestation in Early China: How People Adapted to Wood Scarcity." In *The Cultivated Forest: People and Woodlands in Asian History,* edited by Ian M. Miller, Bradley Camp Davis, Brian Lander, and John S. Lee, 1–19. Seattle: University of Washington Press, 2022.

Latour, Bruno. "Agency at the Time of the Anthropocene." *New Literary History* 45, no. 1 (2014): 1–18.

Laughlin, Charles. *Chinese Reportage: The Aesthetics of Historical Experience.* Durham, NC: Duke University Press, 2002.

Lean, Eugenia. *Public Passions: The Trial of Shi Jianqiao and the Rise of Popular Sympathy in Republican China.* Berkeley: University of California Press, 2007.

Lee, Haiyan. *Revolution of the Heart: A Genealogy of Love in China, 1900–1950.* Stanford, CA: Stanford University Press, 2005.

Leenhouts, Mark. "Culture against Politics: Roots-Seeking Literature." In *The Columbia Companion to Modern Chinese Literature,* edited by Kirk Denton, 299–306. New York, NY: Colombia University Press, 2016.

Li, Changjia, Bojie Fu, Shuai Wang, Lindsay C. Stringer, Yaping Wang, Zidong Li, Yanxu Liu, and Wenxin Zhou. "Drivers and Impacts of Changes in China's Drylands." *Nature Reviews Earth and Environment* 2, no. 12 (2021): 858–73.

Li, Cheng. "Echoes from the Opposite Shore: Chinese Ecocritical Studies as a Transpacific Dialogue Delayed." *Interdisciplinary Studies in Literature and Environment* 21, no. 4 (2014): 821–43.

Li, Cheng, and Yanjun Liu. "Reading Climate Change in the Anthropocene: Material Ecocriticism and Chinese Environmental Literature." *Green Letters* 21, no. 2 (2017): 138–51.

Li, Jie. *Cinematic Guerrillas: Propaganda, Projectionists, and Audiences in Socialist China.* New York, NY: Columbia University Press, 2023.

Li, Lillian M. *Fighting Famine in North China: State, Market, and Environmental Decline, 1690s–1990s.* Stanford, CA: Stanford University Press, 2007.

Li, Meng, and King-fai Tam. "Literature of Trauma and Reflection." In *Routledge Handbook of Modern Chinese Literature,* edited by Ming Dong Gu, 450–61. New York, NY: Routledge, 2019.

Li, Yifei, and Judith Shapiro. *China Goes Green: Coercive Environmentalism for a Troubled Planet.* New York, NY: Polity Press, 2020.

Liu, Jianmei. *Revolution Plus Love: Literary History, Women's Bodies, and the Thematic Repetition in Twentieth-Century Chinese Fiction.* Honolulu: University of Hawai'i Press, 2003.

Liu, Lydia. *Translingual Practice: Literature, National Culture, and Translated Modernity; China, 1900–1937.* Stanford, CA: Stanford University Press, 1995.

Lo, Carlos Wing-Hung, Ning Liu, Xue Pang, and Pansy Hon Ying Li. "Unpacking the Complexity of Environmental Regulatory Governance in a Globalizing World: A Critical Review for Research Agenda Setting." *Journal of Environmental Policy and Planning* 22, no. 5 (2020): 594–607.

Lu, Sheldon, and Haomin Gong, eds. *Ecology and Chinese-Language Cinema: Reimagining a Field.* New York, NY: Routledge, 2019.

Lu, Xiaoning. "The Politics of Recognition and Constructing Socialist Subjectivity: Reexamining the National Minority Film (1949–1966)." *Journal of Contemporary China* 23, no. 86 (2014): 372–86.

Lytle, Mark Hamilton. *The Gentle Subversive: Rachel Carson, Silent Spring, and the Rise of the Environmental Movement.* New York, NY: Oxford University Press, 2007.

Ma, Zhijun, Ying Chen, David S. Melville, Jun Fan, Jianguo Liu, Jinwei Dong, Kun Tan, et al. "Changes in Area and Number of Nature Reserves in China." *Conservation Biology* 33, no. 5 (2019): 1066–75.

Maher, Neil. *Nature's New Deal: The Civilian Conservation Corps and the Roots of the American Environmental Movement.* New York, NY: Oxford University Press, 2008.

Marder, Michael. *Plant-Thinking: A Philosophy of Vegetal Life.* New York, NY: Columbia University Press, 2013.

Marks, Robert. *China: Its Environment and History.* New York, NY: Lexington Books, 2011.

Marsh, George Perkins. *Man and Nature: or, Physical Geography as Modified by Human Action.* 1864. Reprint, Seattle: University of Washington Press, 2003.

Martini, Edwin. *Agent Orange: History, Science, and the Politics of Uncertainty.* Amherst: University of Massachusetts Press, 2012.

McElwee, Pamela. *Forests Are Gold: Trees, People, and Environmental Rule in Vietnam.* Seattle: University of Washington Press, 2016.

McGrath, Jason. *Postsocialist Modernity: Chinese Cinema, Literature, and Criticism in the Market Age.* Stanford, CA: Stanford University Press, 2008.

McNeill, J. R. "Woods and Warfare in World History." *Environmental History* 9, no. 3 (2004): 388–410.

McNeill, J. R., and Peter Engelke. *The Great Acceleration: An Environmental History of the Anthropocene since 1945.* Cambridge, MA: Harvard University Press, 2016.

Meng, Yue. "Unspeakable Ecology: Eco-science and Environmental Awareness through Thick Inquiries, 1910s–1980s." *Twentieth-Century China* 47, no. 2 (2022): 91–111.

Menzies, Nicholas K. *Forest and Land Management in Imperial China.* New York, NY: St. Martin's Press, 1994.

Merkel-Hess, Kate. *The Rural Modern: Reconstructing the Self and State in Republican China*. Chicago: University of Chicago Press, 2016.

Messines, Jean. "Forest Rehabilitation and Soil Conservation in China." *Unasylva* 12, no. 3 (1958): 102–20.

Meyer, John. "Gifford Pinchot, John Muir, and the Boundaries of Politics in American Thought." *Polity* 30, no. 2 (1997): 267–84.

Miller, Char. *Gifford Pinchot and the Making of Modern Environmentalism*. Washington: Island Press, 2013.

Miller, Ian M. *Fir and Empire: The Transformation of Forests in Early Modern China*. Seattle: University of Washington Press, 2020.

———. "Forestry and the Politics of Sustainability in Early China." *Environmental History* 22, no. 4 (2017): 594–617.

Mitchell, Timothy. *Carbon Democracy: Political Power in the Age of Oil*. New York, NY: Verso Books, 2011.

Mitman, Gregg. "Reflections on the Plantationocene: A Conversation with Donna Haraway and Anna Tsing." Edge Effects. Last modified October 12, 2019. https://edgeeffects.net/haraway-tsing-plantationocene/.

Moore, Scott M. "Modernisation, Authoritarianism, and the Environment: The Politics of China's South-North Water Transfer Project." *Environmental Politics* 23, no. 6 (2014): 947–64.

Moratto, Riccardo, ed. *Ecocriticism and Chinese Literature: Imagined Landscapes and Real Lived Spaces*. New York, NY: Routledge, 2022.

Mortimer-Sandilands, Catriona. "Melancholy Natures, Queer Ecologies." In *Queer Ecologies: Sex, Nature, Politics, Desire*, edited by Catriona Mortimer-Sandilands and Bruce Erikson, 331–58. Bloomington: Indiana University Press, 2010.

Mostern, Ruth. *The Yellow River: A Natural and Unnatural History*. New Haven, CT: Yale University Press, 2021.

Mullaney, Thomas. *Coming to Terms with the Nation: Ethnic Classification in Modern China*. Berkeley: University of California Press, 2011.

Muscolino, Micah S. *The Ecology of War in China: Henan Province, the Yellow River, and Beyond, 1938–1950*. New York, NY: Cambridge University Press, 2014.

Musemwa, Muchaparara. "Zimbabwe's Drift toward Authoritarianism and Its Environmental Consequences, 2000–2017." In *Environmentalism under Authoritarian Regimes: Myth, Propaganda, Reality*, edited by Stephen Brain and Viktor Pal, 191–213. New York, NY: Routledge, 2019.

Nash, Roderick. *Wilderness and the American Mind*. 1967. Reprint, New Haven, CT: Yale University Press, 2014.

Nathan, Andrew J. "Authoritarian Resilience." *Journal of Democracy* 14, no. 1 (2003): 6–17.

Nelson, Arvid. *Cold War Ecology: Forests, Farms, and People in the East German Landscape, 1945–1989*. New Haven, CT: Yale University Press, 2005.

"1959." Zhongguokexueyuan 中国科学院 [Chinese Academy of Sciences]. Accessed June 6, 2019. http://www.cas.cn/jzzky/ysss/bns/200909/t20090928_2529131.shtml.

Nixon, Rob. *Slow Violence and the Environmentalism of the Poor*. Cambridge, MA: Harvard University Press, 2011.

Norton, Bryan. "Environmental Ethics and Weak Anthropocentrism." *Environmental Ethics* 6, no. 2 (1984): 131–48.

Noy-Meir, Imanuel. "Desert Ecosystems: Environment and Producers." *Annual Review of Ecology and Systematics* 4, no. 1 (1973): 25–51.

Odum, Eugene P. "The Ecosystem Approach in the Teaching of Ecology Illustrated with Sample Class Data." *Ecology* 38 (1957): 531–35.

———. *Fundamentals of Ecology*. Philadelphia, PA: Saunders, 1953.

Odum, Howard T. *Environment, Power, and Society*. New York, NY: Wiley, 1971.

Ouyang Qin 欧阳勤, Li Zhengfeng 李正峰, and Ouyang Ping 欧阳屏. *Heisedang'an: Teda zainan shijian jishi* 黑色档案：特大灾难事件纪实 [Black archives: documentary of major disasters]. Shenzhen: Haitianchubanshe, 1999.

Pak, Chris. *Terraforming: Ecopolitical Transformations and Environmentalism in Science Fiction*. Liverpool: Liverpool University Press, 2016.

Papacharissi, Zizi. *Affective Publics: Sentiment, Technology, and Politics*. New York, NY: Oxford University Press, 2015.

Parker, Elizabeth. *The Forest and the EcoGothic: The Deep Dark Woods in the Popular Imagination*. New York, NY: Palgrave, 2020.

Pei, Minxin. *China's Trapped Transition: The Limits of Developmental Autocracy*. Cambridge, MA: Harvard University Press, 2006.

Perdue, Peter C. *China Marches West: The Qing Conquest of Central Eurasia*. Cambridge, MA: Harvard University Press, 2009.

———. *Exhausting the Earth: State and Peasant in Hunan, 1500–1850*. Cambridge, MA: Harvard University Asia Center, 1987.

Perry, Elizabeth. *Anyuan: Mining China's Revolutionary Tradition*. Berkeley: University of California Press, 2012.

———. "Moving the Masses: Emotion Work in the Chinese Revolution." *Mobilization: An International Quarterly* 7, no. 2 (2002): 111–28.

Pietz, David. *The Yellow River: The Problem of Water in Modern China*. Cambridge, MA: Harvard University Press, 2015.

Pitts, Larissa. "Unity in the Trees: Arbor Day and Republican China, 1915–1927." *Journal of Modern Chinese History* 13, no. 2 (2019): 296–318.

Pollan, Michael. *The Botany of Desire: A Plant's-Eye View of the World*. New York, NY: Random House, 2001.

———. *Second Nature: A Gardener's Education*. New York, NY: Grove Press, 1991.

Pomeranz, Kenneth. *The Great Divergence: China, Europe, and the Making of the Modern World Economy*. Princeton, NJ: Princeton University Press, 2000.

———. *The Making of a Hinterland: State, Society, and Economy in Inland North China, 1853–1937*. Berkeley: University of California Press, 1993.

Purdy, Jedediah. *After Nature: A Politics for the Anthropocene*. Cambridge, MA: Harvard University Press, 2015.

Radkau, Joachim. *Nature and Power: A Global History of the Environment*. Translated by Thomas Dunlap. New York, NY: Cambridge University Press, 2009.

Rafel, Lisa. *Desiring China: Experiments in Neoliberalism, Sexuality, and Public Culture*. Durham, NC: Duke University Press, 2007.

Reed, Christopher. "Sino-silviculture: State-Sponsored Green Forestry Initiatives in Mao's China." In *Environmentalism under Authoritarian Regimes: Myth, Propaganda, and Reality*, edited by Stephen Brain and Viktor Pal, 180–90. New York, NY: Routledge, 2018.

Reischl, Katherine. *Photographic Literacy: Cameras in the Hands of Russian Authors*. Ithaca, NY: Cornell University Press, 2018.

Richardson, Stanley. *Forests and Forestry in China: Changing Patterns of Resource Development*. Washington, DC: Island Press, 1990.

Rogaski, Ruth. *Hygienic Modernity: The Meanings of Health and Disease in Treaty Port China*. Berkeley: University of California Press, 2004.

———. *Knowing Manchuria: Environments, the Senses, and Natural Knowledge on an Asian Borderland*. Chicago: University of Chicago Press, 2022.

Rojas, Carlos. "Language, Ethnicity, and the Politics of Literary Taxonomy: Ng Kim Chew and Mahua Literature." *PMLA* 131, no. 5 (2016): 1316–27.

Rothstein, Bo, and Jan Teorell. "What Is Quality of Government? A Theory of Impartial Government Institutions." *Governance* 21, no. 2 (2008): 165–90.

Rutkow, Eric. *American Canopy: Trees, Forests, and the Making of a Nation*. New York, NY: Simon and Schuster, 2012.

Schlesinger, Jonathan. *A World Trimmed with Fur: Wild Things, Pristine Places, and the Natural Fringes of Qing Rule*. Stanford, CA: Stanford University Press, 2017.

Schmalzer, Sigrid. *Red Revolution, Green Revolution: Scientific Farming in Socialist China*. Chicago: University of Chicago Press, 2016.

Schneider, Laurence. "Lysenkoism in China: Proceedings of the 1956 Qingdao Genetics Symposium." *Chinese Law and Government* 19, no. 2 (Summer 1986): iii–xxii.

Scott, James C. *The Art of Not Being Governed: An Anarchist History of Upland Southeast Asia*. New Haven, CT: Yale University Press, 2009.

———. *Seeing like a State: How Certain Schemes to Improve the Human Condition Have Failed*. New Haven, CT: Yale University Press, 1998.

Seow, Victor. *Carbon Technocracy: Energy Regimes in Modern East Asia*. Chicago: University of Chicago Press, 2022.

Shaan Gan Ning bianqu caizhengjingjishi bianxiezu 陕甘宁边区财政经济史编写组, ed. *Kangrizhanzhengshiqi Shaan Gan Ning bianqu caizhengjingji shiliaozhaibian* 抗日战争时期陕甘宁边区财政经济史料摘编 [Excerpts from financial and economic historical materials of the Shaanxi-Gansu-Ningxia border region during the Anti-Japanese War]. Vol. 2. Wuhan: Changjiangwenyi chubanshe, 2016.

Shapiro, Judith. *Mao's War against Nature: Politics and the Environment in Revolutionary China*. New York, NY: Cambridge University Press, 2001.

Shearman, D. J. C., and J. W. Smith. *The Climate Change Challenge and the Failure of Democracy*. Westport, CT: Praeger, 2007.

Shen Guofang 沈国舫 and Liu Jia 刘佳. "Zhongguo gaodeng linyejiaoyu de jiegou jiqi tiaozheng" 中国高等林业教育的结构及其调整 [The structure and adjustment of higher forestry education in China]. In *Nongyejiaoyu dexianzhuang he zhanwang guoji taolunhui lunwenji* 农业教育的现状和展望国际讨论会论文集 [Proceedings of the International Symposium on the Current Situation and Prospects of Agricultural Education], 186–93. Beijing: Beijing nongyedaxue chubanshe, 1990.

Shih, Shu-mei. "Introduction: What Is Sinophone Studies?" In *Sinophone Studies: A Critical Reader*, edited by Shu-mei Shih, Chien-hsin Tsai, and Brian Bernards, 1–16. New York, NY: Columbia University Press, 2013.

Sideris, Lisa H., and Kathleen Dean Moore, eds. *Rachel Carson: Legacy and Challenge*. Albany: State University of New York Press, 2008.

Smith, Michael B. "The Value of a Tree: Public Debates of John Muir and Gifford Pinchot." *Historian* 60, no. 4 (1998): 757–78.

So, Alvin. "The Rise of Authoritarianism in China in the Early 21st Century." *International Review of Modern Sociology* 45, no. 1 (2019): 49–70.

Song, Mingwei. *Fear of Seeing: A Poetics of Chinese Science Fiction*. New York, NY: Columbia University Press, 2023.

———. *Young China: National Rejuvenation and the Bildungsroman, 1900–1959*. Cambridge, MA: Harvard University Asia Center, 2015.

Song, Weijie. "Environmentality, Sustainability, and Chinese Storytelling." *Cultura* 20, no. 1 (2023): 55–66.

Songster, E. Elena. *Panda Nation: The Construction and Conservation of China's Modern Icon*. New York, NY: Oxford University Press, 2018.

Stoma, Monika, and Agnieszka Dudziak. "Analysis of Market Behaviour on the Organic Food Market in Terms of Environmental Protection and Consumer Environmentalism." In *International Symposium on Farm Machinery and Processes Management in Sustainable Agriculture*, edited by Simone Pascuzzi and Francesco Santoro, 345–56. Cham: Springer, 2022.

Strand, David. *An Unfinished Republic: Leading by Word and Deed in Modern China*. Berkeley: University of California Press, 2011.

Stross, Randall E. *The Stubborn Earth: American Agriculturalists on Chinese Soil, 1898–1937*. Berkeley: University of California Press, 1986.

Sun Hui 孙辉 and Wang Jinjun 王进举. "Senlin weishi—Wujing senlinbudui de zhandoulicheng he fazhanyange" 森林卫士—武警森林部队的战斗历程和发展沿革 [Forest guardians—the combat history and development history of the People's Armed Police Forestry Force]. *Junshilishi* 军事历史, no. 8 (2006): 60–64.

Taglicozzo, Eric, Helen F. Siu, and Peter C. Perdue, eds. *Asia Inside Out: Connected Places*. Cambridge, MA: Harvard University Press, 2015.

Tang, Xiaobing. *Visual Culture in Contemporary China*. Cambridge: Cambridge University Press, 2015.

Tansley, Arthur G. "The Use and Abuse of Vegetational Concepts and Terms." *Ecology* 16, no. 3 (1935): 284–307.

Thornber, Karen. *Ecoambiguity: Environmental Crises and East Asian Literatures*. Ann Arbor: University of Michigan Press, 2012.

Totman, Conrad. *The Green Archipelago: Forestry in Preindustrial Japan*. Berkeley: University of California Press, 1989.

Tsing, Anna. *Friction: An Ethnography of Global Connection*. Princeton, NJ: Princeton University Press, 2005.

Tuan, Yi-fu. *Topophilia: A Study of Environmental Perceptions, Attitudes, and Values*. New York, NY: Columbia University Press, 1974.

Tyrrell, Ian. *Crisis of the Wasteful Nation: Empire and Conservation in Theodore Roosevelt's America*. Chicago: University of Chicago Press, 2015.

Uekötter, Frank. *The Green and the Brown: A History of Conservation in Nazi Germany*. New York, NY: Cambridge University Press, 2006.

United Nations. "The Paris Agreement." Accessed March 22, 2023. https://unfccc.int/process-and-meetings/the-paris-agreement/the-paris-agreement.

Usner, Daniel H. "Borderlands." In *A Companion to Colonial America*, edited by Daniel Vickers, 408–24. New York, NY: Blackwell, 2003.

Visser, Robin. *Questioning Borders: Ecoliteratures of China and Taiwan*. New York, NY: Columbia University Press, 2023.

Volland, Nicolai. *Socialist Cosmopolitanism: The Chinese Literary Universe, 1945–1965*. New York, NY: Columbia University Press, 2017.

Walker, Brett. *The Lost Wolves of Japan*. Seattle: University of Washington Press, 2005.

Wallace, Vesna A. "Buddhist Sacred Mountains, Auspicious Landscapes, and Their Agency." In *Buddhism in Mongolian History, Culture, and Society*, edited by Vesna A. Wallace, 221–40. Oxford: Oxford University Press, 2015.

Wang, Ban. *At Home in Nature: Technology, Labor, and Critical Ecology in Modern China*. Durham, NC: Duke University Press, 2023.

———. *The Sublime Figure of History: Aesthetics and Politics in Twentieth-Century China*. Stanford, CA: Stanford University Press, 1997.

Wang, Jing. *High Culture Fever: Politics, Aesthetics, and Ideology in Deng's China*. Berkeley: University of California Press, 1996.

———. *The Other Digital China: Nonconfrontational Activism on the Social Web*. Cambridge, MA: Harvard University Press, 2019.

Wang Nuo 王诺. *Shengtaipipingyu shengtai sixiang* 生态批评与生态思想 [Ecological thought and ecocriticism]. Beijing: Renmin chubanshe, 2013.

Wang Xiliang 王希亮. "Jindai zhongguodongbei senlin de zhiminkaifa yu shengcunkongjian bianqian" 近代中国东北森林的殖民开发与生态空间变迁 [Colonial development and ecological spatial changes in forests in northeastern China in modern times]. *Lishiyanjiu* 历史研究1 (2017): 85–103.

Wang Xiqun 王希群, Jiang Zeping 江泽平, Wang Anqi 王安琪, Guo Baoxiang 郭保香, and Yang Shaolong 杨绍陇, eds. *Zhongguo linye shiye de xianqu he kaituozhe* 中国林业事业的先驱和开拓者 [The pioneers of Chinese forestry]. Beijing: Zhongguo linye chubanshe, 2018.

Wang, Yuke, Rui Jiang, MeiJu Jiao, Tiancai Cao, and Xiang Yu. "Macro and Micro Experimental Study on Solidification of Yellow River Silt Based on Different Biomineralization Technologies." *Environmental Earth Sciences* 82, no. 3 (2023): 1–17.

Warde, Paul, Libby Robin, and Sörlin Sverker. *The Environment: A History of the Idea*. Baltimore: Johns Hopkins University Press, 2018.

Weiner, Benno. *The Chinese Revolution on the Tibetan Frontier*. Ithaca, NY: Cornell University Press, 2020.

Weiner, Douglas R. *Models of Nature: Ecology, Conservation, and Cultural Revolution in Soviet Russia*. Bloomington: Indiana University Press, 2000.

Weller, Robert P., and Peter K. Bol. "From Heaven-and-Earth to Nature: Chinese Concepts of the Environment and Their Influence on Policy Implementation." In *Confucianism and Ecology: The Interrelation of Heaven, Earth, and Humans*, edited by Mary Tucker and John Berthrong, 313–42. Cambridge, MA: Harvard University Press, 1998.

Wells, Audrey. *The Political Thought of Sun Yat-sen: Development and Impact*. New York, NY: Palgrave, 2001.

Widick, Richard. *Trouble in the Forest: California's Redwood Timber Wars*. Minneapolis: University of Minnesota Press, 2009.

Williams, Dee Mack. *Beyond Great Walls: Environment, Identity, and Development on the Chinese Grasslands of Inner Mongolia*. Stanford, CA: Stanford University Press, 2002.

Williams, Raymond. *Marxism and Literature*. New York, NY: Oxford University Press, 1978.

Wilson, Edward O. "Biophilia and the Conservation Rhetoric." In *The Biophilia*

Hypothesis, edited by Stephen R. Kellert and Edward O. Wilson, 31–41. Washington, DC: Island Press, 1995.

Wilson, Jeffrey K. *The German Forest: Nature, Identity, and the Contestation of a National Symbol, 1871–1914*. Toronto: University of Toronto Press, 2012.

Wohlleben, Peter. *The Hidden Life of Trees: What They Feel, How They Communicate; Discoveries from a Secret World*. Translated by Jane Billinghurst. Vancouver: Greystone Books, 2016.

Worster, Donald. *Nature's Economy: A History of Ecological Ideas*. Cambridge: Cambridge University Press, 1994.

———. *A Passion for Nature: The Life of John Muir*. New York, NY: Oxford University Press, 2008.

Wu Jinzan 吴金赞. *Zhonghua minguo linye fazhishi* 中华民国林业法制史 [Forestry legal system of the Republic of China]. Taipei: Zhengzhong shuju, 1991.

Wu, Shellen. *Empires of Coal: Fueling China's Entry into the Modern World Order, 1860–1920*. Stanford, CA: Stanford University Press, 2015.

Xiao, Tie. *Revolutionary Waves: The Crowd in Modern China*. Cambridge, MA: Harvard University Press, 2017.

Xiong Datong 熊大桐. *Zhongguo jindai linye shi* 中国近代林业史 [Modern forestry history in China]. Beijing: Zhongguo linye chubanshe, 1994.

Xu, Yunan. "Land Grabbing by Villagers? Insights from Intimate Land Grabbing in the Rise of Industrial Tree Plantation Sector in Guangxi, China." *Geoforum* 96, no. 4 (2018): 141–49.

Yang, Guobin, and Craig Calhoun. "Media, Civil Society, and the Rise of a Green Public Sphere in China." *China Information* 21, no. 2 (July 2007): 211–36.

Yang Shaozhang 杨绍章 and Xin Yejiang 辛业江. *Zhongguo linye jiaoyushi* 中国林业教育史 [History of forestry education in China]. Beijing: Zhongguolinye chubanshe, 1988.

Yao Dan 姚丹. *Geming zhongguo de tongsu tezheng yu zhuti jiangou: "Linhaixueyuan" jiqi yanshengwenben kaocha* 革命中国的通俗特征与主体建构:《林海雪原》及其衍生文本考察 [Popular characteristics and main body construction of revolutionary China: a study of *Tracks in the snowy forest* and its derived texts]. Beijing: Beijing daxue chubanshe, 2011.

Yasutomi Ayumu 安冨歩 and Fukao Yōko 深尾葉子, eds. *"Manshū" no seiritsu: Shinrin no shōjin to kindai kūkan no keisei* "満洲"の成立: 森林の消尽と近代空間の形成 [The formation of "Manchuria": forest depletion and the creation of modern spaces]. Nagoya-shi: Nagoya daigaku shuppankai, 2009.

Yeh, Emily T. "The Making of Natural Infrastructure in China's Era of Ecological Civilization." *China Quarterly* 255 (September 2023): 611–27.

Yin, Zun, Catherine Ottlé, Philippe Ciais, Feng Zhou, Xuhui Wang, Polcher Jan, Patrice Dumas, et al. "Irrigation, Damming, and Streamflow Fluctuations of the Yellow River." *Hydrology and Earth System Sciences* 25, no. 3 (2021): 1133–50.

Yue, Gang. "The Strange Landscape of the Ancient: Environmental Consciousness in '*The King of Trees.*'" *American Journal of Chinese Studies* 5, no. 1 (1998): 68–88.

Zanasi, Margherita. *Saving the Nation: Economic Modernity in Republican China*. Chicago: University of Chicago Press, 2006.

Zhai, Junjie, Ling Wang, Yuan Liu, Chengyuan Wang, and Xuegang Mao. "Assessing the Effects of China's Three-North Shelter Forest Program over 40 Years." *Science of the Total Environment* 857 (2023): 1–12.

Zhang Lianwei 张连伟, Li Fei 李飞, and Zhou Jingyong 周景勇, eds. *Zhongguo gudai linyewenxian xuandu* 中国古代林业文献选读 [Selected readings from ancient Chinese forestry literature]. Beijing: Zhongguo yanshan chubanshe, 2015.

Zhang, Ling. *The River, the Plain, and the State: An Environmental Drama in Northern Song China, 1048–1128*. New York, NY: Cambridge University Press, 2016.

Zhang, Meng. *Timber and Forestry in Qing China: Sustaining the Market*. Seattle: University of Washington Press, 2021.

Zhao, Dingxin. *The Power of Tiananmen: State-Society Relations and the 1989 Beijing Student Movement*. Chicago: University of Chicago Press, 2004.

Zheng Yu 郑宇. *Jindai dongbei senlin ziyuan chanyehua yanjiu (1878–1931 nian)* 近代东北森林资源产业化研究 (1878–1931年) [Modern northeastern forestry resource industrialization study (1878–1931)]. Shanghai: Shanghai shehuikexue chubanshe, 2020.

Zhu, Annah Lake. *Rosewood: Endangered Species Conservation and the Rise of Global China*. Cambridge, MA: Harvard University Press, 2022.

Zhu Yu 朱羽. *Shehuizhuyi yu "ziran": 1950–1960 niandai Zhongguo meixue lunzheng yu wenyi shijian yanjiu* 社会主义与"自然"：1950–1960年代中国美学论争与文艺实践研究 [Socialism and "nature": study of aesthetic debates and artistic practices in China during the 1950s–1960s]. Beijing: Beijing daxue chubanshe, 2020.

INDEX

MIX
Papier | Fördert
gute Waldnutzung
FSC® C083411

Kolibri 360 GmbH
Ferdinand-Jühlke-Straße 7
99095 Erfurt, Deutschland
produktsicherheit@kolibri360.de